# Beginning
## Shepherd's Manual
### SECOND EDITION

# Beginning Shepherd's Manual

## SECOND EDITION

Barbara Smith, MS
Mark Aseltine, PhD
Gerald Kennedy, DVM

Iowa State University Press / Ames

**Barbara Smith** received her BA in English and MS in educational psychology. She is a journalist by training, writing extensively on agricultural and medical topics, and raised sheep for 10 years in Solon, Iowa.

**Mark Aseltine** received his PhD from Oregon State University in Corvallis. His research includes work on both sheep and dairy nutrition, and he is an independent ruminant and management consultant.

**Gerald Kennedy** received his DVM from the College of Veterinary Medicine, Iowa State University, Ames, and is currently a partner in a general practice. He has raised both purebred and commercial sheep and has been a consultant in health and nutrition, management, and production of sheep on a national and international basis.

© 1997 Iowa State University Press, Ames, Iowa 50014
All rights reserved

Orders: 1-800-862-6657        Fax:        1-515-292-3348
Office:  1-515-292-0140        Web site: www.isupress.edu

Authorization to photocopy items for internal or personal use, or the internal or personal use of specific clients, is granted by Iowa State University Press, provided that the base fee of $.10 per copy is paid directly to the Copyright Clearance Center, 222 Rosewood Drive, Danvers, MA 01923. For those organizations that have been granted a photocopy license by CCC, a separate system of payments has been arranged. The fee code for users of the Transactional Reporting Service is 0-8138-2799-X/97 $.10.

♾ Printed on acid-free paper in the United States of America
First edition, 1983
Second edition, 1997

Cover photo of the late Everett Whiting by Mark Lovewell.

Library of Congress Cataloging-in-Publication Data

Smith, Barbara
    Beginning shepherd's manual / Barbara Smith, Mark Aseltine, Gerald Kennedy.—
2nd ed.
        p.        cm.
    Includes bibliographical references (p.        ) and index.
    ISBN 0-8138-2799-X
    1. Sheep—Handbooks, manuals, etc. I. Aseltine, Mark. II. Kennedy, Gerald.
III. Title.
SF375.S59        1997
636.3—dc21                                                                96-51981

Last digit is the print number:  9  8  7  6  5  4  3  2

# CONTENTS

Thanks to Jon

# ACKNOWLEDGMENTS

As manager of this project, I want to first acknowledge my continued indebtedness to Tom Wickersham, retired Extension Sheep Specialist at Iowa State University. Tom was the original inspiration for the first edition with his gentle and nonjudgmental approach to teaching and his love of sheep. He is a true shepherd. I also extend continued thanks to Keith Miller, DVM, who was the veterinary consultant on the first edition. Dan Morrical of Iowa State University, Dave Thomas of the University of Wisconsin, and Lyle McNeal of Utah State University were of great assistance in updating the book. The Iowa State University Extension Service has been a generous and ever ready resource of information and materials.

Carolyn Milligan, the artist and designer of the first edition, provided new illustrations for this edition, as well as serving as a wonderful resource for my "sheep tour" of the southwest. For anyone visiting Gallup, New Mexico, the sculpture she designed and worked on in the courthouse square is worth a visit. It represents a Navajo weaving, a wonderful tribute to the history of the area and even to the role of sheep in it. I also thank design artist Lee Previant for his help with updating various graphics.

Dr. Mark Aseltine, who wrote the nutrition chapter for this edition, wishes to thank Lori Gerhardt for typing the chapter, as well as his family, wife Elena and children Heather, Leah, and Aaron, for their nurturance.

Dr. Gerald Kennedy, of the Pipestone Veterinary Clinic in Pipestone, Minnesota, acknowledges the contributions of Dr. J.D. Bobb, G.D. Spronk, Barry Kerkaert, and the staff of the Lamb and Wool Program, Southwestern Technical College. He also wishes to express his gratitude for the longtime support of his wife, Kay, who passed away in October 1996.

Suzanne Erenberger did the majority of the word processing for this edition, dealing with my difficult handwriting and cutting and pasting, as well as interpreting the handwriting of other contributors. Many thanks for her patience, skill, promptness, and good humor.

And special thanks to the many good shepherds who allowed me to visit them in the process of putting together this new edition. As they meet new challenges in their industry, they continue to honorably represent a wonderful and often misunderstood segment of American agriculture.

*Barbara Smith*

# Beginning
## Shepherd's Manual

SECOND EDITION

# 1. The Beginning Shepherd

*"The sheep have always been with us. The spirit of the sheep has always been here."*—Joe Benally

*"I'm still a shepherd. That's what I like best."*
—Spence Rule

Becoming a shepherd means taking on an ancient tradition in a modern setting that is not always hospitable. Once a major part of American agriculture, sheep production today is considered a relatively minor industry, and this presents many frustrations for both new and experienced sheep producers. "The biological beauty of the species is still there," says Dr. Charles Parker of the American Sheep Industry Association, "But you look at the numbers and you say 'What's wrong?' It's not the products. It's not the biology of the sheep ... they can compete." Health care, distribution systems, and marketing are all affected by the relatively small numbers. Even learning the shepherd's work is affected.

But the choice to raise sheep goes beyond an interest in business or even farming. Shepherds have a respect and affection for the animal, in spite of what the winds of agricultural fashion and income bring.

The beginning shepherd can start with sudden immersion. Just a good idea. A livestock truck pulls into the yard and deposits 40 sheep in a small, poorly fenced pasture. Rumors are that these animals are not intelligent, but within 15 minutes they are out of the assigned space and wandering down a dirt road looking for choicer ground, making anxious noises as they trot.

Learning about sheep takes place as problems arise—footrot, overeating, a rash of dead lambs. Crisis learning is a powerful method of education, but it can be hard on the sheep and the shepherd. While it isn't possible to avoid all beginner trauma, it can be minimized. Just as there is a system for shearing that results in the best fleece and the most comfortable shearer, there is also a well-established plan for taking care of sheep.

The basics of the system are ancient, although certainly the specifics have changed with technology and with the intensity of the operations. *The Practical Shepherd,* written in 1848 by veterinarian Henry Randall, uses the same outline as this book, and the overall management approach is very similar.

The purpose of this book is to help provide an orderly approach to learning for the first-time shepherd. Panic and confusion can hide many of the wonderful things about raising sheep. Being a shepherd provides moments of joy quite unrelated to producing a crop of good market lambs. Sheep make good family projects because they can be handled easily, even by children, taking more patience and psychology than force.

Sheep are famous for their lack of aggression. Even in the most hectic times, you may be struck by scenes from your daily sheep chores—watching month-old lambs play king of the mountain while their mothers graze together; a child followed by a chunky bottle lamb crossing a dandelion-spattered pasture. And there is the wonder of discovering a ewe in labor on a cold still night. She has one lamb and then another. She licks their steaming wet bodies and chuckles to them until they stagger to their feet and find their way to that first lifesaving milk. You may sit and watch, doing exactly what shepherds have done for thousands of years ... bear concerned witness to a great miracle of birth and survival.

## BASIC FACTS ABOUT SHEEP

The shepherd looks after *Ovis ammon aries,* the domestic sheep (Fig. 1.1.).

The sheep is a member of the bovine family of hollow-horned ruminants. The two genera, goats (*Capra*) and sheep (*Ovis*), are closely related. The chief difference is that goats have a divided upper lip and sheep do not. Male sheep do not have a tuft under their chins as do billy goats. Most sheep have distinctive glands that open on the surface of the skin below the eyes and secrete an oily substance. Sheep also have a gland between their toes, but its purpose is not really known. Some experts say the gland helps wandering sheep find their flock, while others say it secretes waste products. One false rumor is that the odor from the secretions will keep cattle from grazing where sheep have stepped. The skin of sheep is

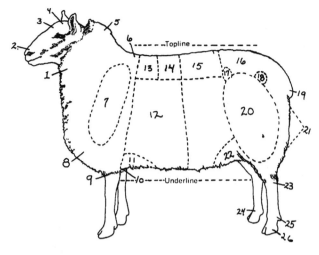

**1.1.** Anatomy of a sheep.

| | | |
|---|---|---|
| 1. Dewlap | 10. Heart | 19. Dock or tail |
| 2. Muzzle | 11. Foreflank | 20. Leg of mutton |
| 3. Forehead | 12. Ribs | 21. Twist |
| 4. Poll | 13. Rack | 22. Rear flank |
| 5. Crest of the neck | 14. Back | 23. Hock |
| 6. Withers | 15. Loin | 24. Pastern |
| 7. Shoulder | 16. Rump | 25. Dewclaw |
| 8. Breast or brisket | 17. Hook bone | 26. Foot |
| 9. Forearm or shank | 18. Pin bone | |

much thinner than the skin of other members of the bovine family, making it especially important to handle sheep carefully.

Sheep have a distinctive behavioral characteristic of flocking. The flocking instinct makes the shepherd's vigilance critical, but at the same time makes sheep relatively easy to handle. The instinct may have evolved because sheep with "solitary" traits were not able to defend themselves against predators and did not survive well—those that did survive were those that kept together under a shepherd's care. The strength of this instinct varies with breeds. The Churro and Rambouillet, for example, have strong instincts compared with the Suffolk and Hampshire.

Sheep also have a tendency to move toward light and are easily stressed by noise, all important in handling and moving them. With an understanding of sheep behavior, it is quite possible to set up a sheep area where one person alone can move several hundred sheep.

The simple watchfulness of the traditional shepherd is still critical to the successful sheep operation. Because of the minor role that sheep play in the U.S. farm economy, it is sometimes hard to learn the details of the shepherd's job. This is not a new problem and is described well in an early twentieth century book on sheep production:

> In America, sheep farming is little understood. Sheep are kept in a more or less desultory manner, having the run of some hill pasture or woodland, fed at intervals in winter, sold off when prices become low, bought up again with the return of higher prices, given small care or encouragement, often afflicted with parasites, internal and external, a side issue with the farmer, profitable in spite of his neglect, yet not often assuming the dignity of a business of themselves. There are several reasons for this state. It is in part a heritage of the days when sheep were little valued for their flesh and were kept mainly for their fleeces. It is in part a result of our once cheap lands and insufficient labor with which to till them. And in large part it is because of ignorance about profitable methods. When sheep thrive, their owners gladly reap the profits; when they become diseased and unprofitable, it is usually charged to "bad luck." There need be a small element of luck or chance in sheep management. There is always a reason for the thriftiness or unthriftiness of a flock. A healthy sheep is certain to be a profitable one.(1)

Because of the relative lack of knowledge about sheep in this country, there are many prejudices and misunderstandings that the beginning shepherd will need to be aware of and sift through. "A sick sheep is a dead sheep" or "There must be money in 'em, cause you can't get it out of 'em" come from a lack of attention to the animal.

Beginning shepherds have three main resources for their own education: professional help, through veterinarians and extension service personnel; consulting with other sheep producers; and written materials.

Specifics about veterinary services will be covered in Chapter 7. It is important to establish a good working relationship with a veterinarian early in your efforts. State extension services are excellent starting points for information. If your local office does not have a sheep expert on hand, it can often direct you to reputable area producers. State offices, as well as local ones, are available for consultation by phone and, now, often through e-mail as well. The extension personnel will not only have information about sheep in general, but they will have a sense of local health problems and feed availability that will be useful. Clinics and field days, which are wonderful opportunities to learn and meet other sheep producers, are often organized through these offices.

Skepticism about books as a source of practical farm information is common. You just have to remember that some printed material about sheep will be helpful to your operation and some will not. With written materials you have much greater access to opinions, research, and techniques than you do through face-to-face conversations. *The Sheep Production Handbook* (2), for example, is an excellent resource  for all aspects of shepherding.

However, comfort and inspiration as well as practical information can come from other sheep producers. People who raise sheep are not afraid to tell stories about themselves. If you have just lost a favorite ewe in a difficult lambing or had a lamb ready for market die inexplicably, there is nothing more heartening than to hear another shepherd say, "I understand how you feel. The same thing happened to us." Some beginning shepherds have actually apprenticed themselves in an ongoing operation. You can help during a lambing period or volunteer to help with health routines. Your extension office can put you in touch with local producers, if you don't know any.

You are your own best resource in learning about sheep. Becoming the scientist-shepherd is a good goal to have as a beginner. Learn to ask questions, even the most basic ones, without embarrassment. Get answers and the reasons behind them so you can make comparisons when you get different ones, which you will. Second, learn to keep accurate records of what you do. Even in a small operation, they will help you learn what works and what doesn't. Records may involve a formal account of breeding and lambing or an informal diary or calendar of how you handled specific problems and of what does and doesn't work for you. Third, don't be afraid to take action. While you will undoubtedly make mistakes as a beginner, the biggest general mistake of shepherding is letting things slide.

## THE HISTORY OF SHEEP

Sheep need shepherds and shepherds need sheep. Sheep have always needed protection from predators. They have also needed the vigilance of a shepherd because their strong flocking instincts, even in the face of illness and injury, can conceal health problems. In return, sheep have offered humans products important to survival, and they are hardy and adaptable to many environments. Sheep can be produced in climates, on terrains, and under feed conditions where many other animals could not survive, let alone thrive. Breeds adapted to local conditions exist all over the world today.

Sheep were among the first animals to be domesticated, perhaps

8,000 years ago. At first only the skins and meat were valued, but as humans learned to spin and weave the wool, the value of sheep greatly increased. We know that woven wool garments were worn in Babylon (which means the land of wool) as early as 4000 BC.(3) The strong relationship between humans and sheep is clear in both the Old and New Testaments of the Bible, where the watchful shepherd and the dependent sheep take on religious significance.

Domestic sheep arrived in the New World with the early Spanish explorers Columbus, Cortez, and later, Coronado. The first English sheep came over with the settlers of the Jamestown Colony in 1607 and were valued for wool rather than meat. (4) In fact, the development of trade in wool and fabric in the American colonies contributed to the growing tension with the English, who wanted to maintain control of the textile industry. British attempts to suppress the growth of wool production and manufacturing became a focus of revolutionary interest, and in protest, even those who could afford the more expensive English fabrics wore coarse colonial homespuns. The "on location" stories from Massachusetts and the Southwest in Chapter 11 in this book illustrate the history of American sheep in the eastern and western regions.

With westward expansion during the 1800s, the sheep population moved on too. In 1810, most sheep in the United States were located in New England and New York, with the notable exception of those raised by the Navajo in New Mexico. By 1820, the center of the sheep population was in the Ohio valley and Great Lakes area. As cheaper rangelands opened in the West, the bulk of the sheep population shifted, and today most of the nation's sheep are located in the western states.(5)

The focus of the sheep industry in the United States has changed dramatically since its beginnings. It has shifted from wool to meat as its main product, and from east to west geographically. The overall importance of sheep in the agricultural economy is not great. Sheep production has declined steadily from a high of 56 million in 1942 to a low of 8.5 million in 1996.(6) By comparison, production of cattle, hogs, fish, and chicken has risen. This is despite the fact that, most of the time, prices for sheep producers have been comparable to, if not higher than, other livestock prices (Fig. 1.2).

There are a number of reasons why the sheep industry has become smaller. The farm population as a whole has decreased dramatically since World War II. There are fewer but larger farms and less farm labor available. For maximum profits, sheep require intensive labor during lambing to save lambs born in the cold outdoors and to help ewes with difficult

**Livestock Prices**

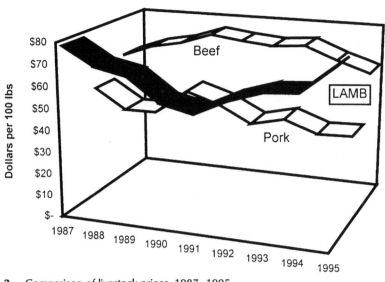

**1.2.**    Comparison of livestock prices, 1987–1995.
(*Courtesy, USDA National Agricultural Statistics Service*)

births. Labor requirement during this time can be eased but not eliminated with modern housing and equipment. Furthermore, predators—primarily coyotes—are a serious and unsolved problem for the western sheep rancher. Finally, many of the lamb slaughterhouses have closed, making marketing more costly. And the palate of the U.S. consumer continues to be a problem.

## THE SHEEP BUSINESS

What does the beginning shepherd need to know about the sheep business? Farming as a whole is a high-risk business. Sheep production seems to have an especially varied history. Sheep numbers are at an all-time low of about 8.5 million. Is this good or bad for the producer? A small but profitable industry is seen as an advantage by some, but most others argue that a greater need for the products needs to be promoted and the industry needs to expand to survive.

In the wool markets, inexpensive, washable synthetic fabrics were developed during World War II and helped to bring wool prices to the point where, in some years, small-flock owners had to pay more for the shearer than the wool was worth. The federal wool incentive payments made wool a more profitable venture, and the 1974 oil crisis made petroleum-based synthetic fabrics less desirable. Washable 100% wool and wool-blend fabrics are now as easy to care for as synthetic fabrics. Natural fabrics have taken on a new popularity over the past decades, and the warmth and comfort of wool are being rediscovered. With the end of the wool incentive payment in 1995, the profitability of wool production will be uncertain. Many producers are seeing this uncertainty as a challenge to create a more valuable product. Others believe that U.S. producers won't be able to survive foreign competition without the subsidies.

Like the popularity of wool, the popularity of lamb has also varied. Lamb has never played the prominent role in American diets that it has in other cultures, especially in the Middle East and Europe. In 1963, Henry Randall wrote:

> Thirty to forty years ago, little mutton was consumed in the United States. Our people had not learned to eat it. Colonizing a new country covered with forests containing animals that prey on sheep, and in which the necessary labor for guarding them was scarce and high, our forefathers kept only enough to meet pressing wants for wool for household use. Beef and pork were more easily grown and better relished. The state of things continued until mutton became a stranger to American tables. The prejudice continued until the comparatively recent introduction of the improved English mutton sheep and until fashion in cities, for once, inaugurated a great and useful change in public taste. Some of the earlier prejudices linger yet in the rural population.(5)

A prejudice against lamb lingers in this country. Per capita consumption of lamb has reached a low of 0.9 pound—a very small portion of the total meat consumption in the country (Fig. 1.3).

The reasons for the low consumption of lamb are a bit circuitous. A trip to the meat department of the local supermarket will explain part of it. Lamb is expensive. It is difficult to develop a new taste in the American consumer when the prices is so high. The price is high, on the other hand, because there isn't very much of it around, because there isn't a demand for it.

Taste and tradition are both factors in the low consumption of lamb. People seem to either love lamb or hate it. Stories are still shared about the awful piece of lamb someone's uncle had to eat in England during

## Consumption per Capita: 1995

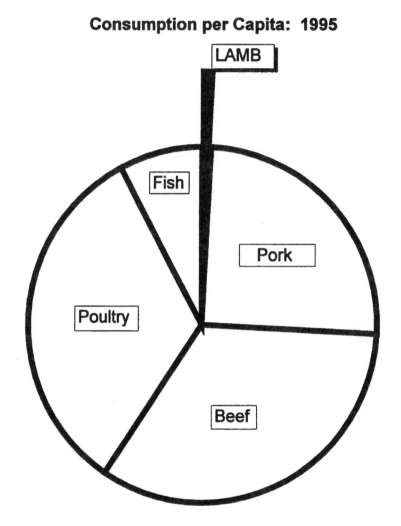

**1.3.**   Per capita meat consumption, in pounds, 1995: lamb, 0.9; pork, 49.5; beef, 63.8; poultry, 63.7; fish, including shellfish, 14.9.
(*Courtesy, USDA Economic Research Service*)

World War II, or lamb that tastes like wool, or lamb that has to be eaten before the fat gets hard. Some of the bad press comes from experiences with poorly cooked lamb, a result of unfamiliarity with cooking it. Leg roasts and chops are the most familiar cuts, and the most expensive. The less expensive cuts such as lamburger or ribs are not common in American kitchens. One butcher explains why he doesn't buy much lamb, although he does get requests for it: "The roasts and chops are gone the first day," he says, "but the rest of it hangs around forever." As a result, the gap between what the producer receives and what the consumer pays is the highest in the meat industry.

Both as individuals and as organizations, sheep producers are looking at ways to cultivate a taste for lamb.

## TYPES OF SHEEP OPERATIONS

### Range Band Flocks

Although the range band flock is not what most beginners start with, this kind of sheep production accounts for about 82% of the sheep in the United States. All or parts of the following states are considered to be range band areas: Arizona, California, Colorado, Idaho, Kansas, Montana, Nevada, New Mexico, North Dakota, Oklahoma, Oregon, South Dakota, Texas, Utah, Washington, and Wyoming.(7) In the range band flocks, 1,000 to 1,500 ewes are usually tended by one or two full-time shepherds. The organization of the band depends on the feed and grazing land available. If there is good pasture, lambs may be raised completely on their mothers' milk and grass. In areas where the forage is poor, lambs may be sold as feeder lambs to be fattened elsewhere. Range bands also supply replacement ewes for flocks in other areas of the country.

Range sheep production depends on large areas of relatively inexpensive pastures, as well as on professional full-time shepherds. Sometimes these operations have sheds where ewes can have lambs indoors in severe weather. In other operations, the ewes lamb on pasture, with a higher death rate but a lower investment in facilities. Because the sheep used in range bands are usually of a heavy wool–producing type, wool is a greater source of income than in other operations.

### Farm Flocks

Farm flocks are usually part of a diversified farming operation. In areas such as the Midwest, where farmers can raise almost any crop or livestock on their land, sheep are often found on farms with acres that cannot or should not be tilled. Meat production is the main income from the farm flock, which varies from small, weed cleanup bands of 10 animals to flocks of several hundred ewes. Farm flocks normally do not require a full-time shepherd. Lambing is scheduled to coincide with the lighter farm workload in the winter and early spring. Because the farmers who own these flocks are usually raising grain crops, almost all lambs are grown to market weight on the farm.

### Purebred Operations

Purebred sheep production is highly specialized. The purebred breeder is in business primarily to provide good breeding stock for commercial sheep producers who, for example, may want to improve growth rate or wool production, and for the show and club lamb industry. Breeding purebred sheep is a tough business for the real beginner, because purebreds are expensive animals to make mistakes on, and the complexities of the purebred market can take considerable time to learn. However, purebred production can be profitable.

Breeders of purebreds usually have their flocks lamb in January, so the ewe and ram lambs will be ready for breeding in the fall. Ewe lambs born later in the year may not breed until the following year. Lambs not meeting breed standards are fed and sold for slaughter. Feed requirements are roughly the same as for other flocks. Most sheep producers, however, place their ewes on maintenance feed through the summer. The purebred breeder may choose to provide more than a maintenance diet so the ewes will look their best for potential buyers during the summer.

### Hobby Sheep

A small, hobby flock can provide owners with meat for the family, wool for spinning or weaving, or energy-efficient lawn mowers on acreages. Extra lambs can be sold to help pay for expenses. Many families make 4-H and purebred shows a family activity. The shepherd who raises sheep as a hobby should be forewarned that more than a few serious full-time producers started with a couple of lambs for the children or for cleaning up weeds around their buildings.

## THE BASIC NEEDS OF SHEEP OPERATIONS

Before considering a particular number of sheep or type of sheep for your operation, you should have some idea of what the labor, feed, and shelter requirements are for the flock. You should also have a rough idea of what your basic expenses and income will be. Shelter, feed, and pasture requirements are discussed in detail in Chapters 3 and 4.

### Labor

The labor required for a flock of 100 ewes is approximately 5 to 7 hours per year per ewe.(8) The heaviest demand for labor comes at lambing time. Table 1.1 shows the number of hours of labor required on a monthly basis for a 100-ewe flock in Wisconsin for February lambing.

The hours of labor don't tell the whole story. Lambing time, even in a small operation, can be a tense period of concentrated work. The shep-

**Table 1.1.   Labor requirements for a 100-ewe flock, February lambing**

|  | Jan | Feb | May | Apr | May | June | July | Aug | Sept | Oct | Nov | Dec |
|---|---|---|---|---|---|---|---|---|---|---|---|---|
| Labor hours | 60 | 150 | 50 | 30 | 50 | 50 | 20 | 15 | 5 | 5 | 5 | 10 |

Courtesy Richard Vatthauer, Extension Livestock Specialist, University of Wisconsin.

herd, particularly in cold climates, gets up twice each night to check the ewes, even in the worst weather. This checking may go on for 30 to 35 days. Most sheep producers agree that this kind of vigilance is essential and can make the difference between a profitable and unprofitable operation.

## Budget

The beginning shepherd has two financial concerns: (1) how much money is needed to buy the sheep and essential equipment and to meet operating expenses (feed, veterinarians etc.) and (2) how much income can be reasonably expected.

The best way to estimate the cost of ewes is to contact your extension

**Table 1.2.   Estimates: Expense-income factors**

|  | Sample | Yours |
|---|---|---|
| A = Number of sheep in flock (4 ewes/acre) | 40 | _____ |
| B = Market price for slaughter lambs/100 lb | $60.00 | _____ |
| C = Market price for wool/lb | .80 | _____ |
| D = Price of corn/bu | 2.50 | _____ |
| E = Price of hay/ton | 60.00 | _____ |
| F = Price of protein supplement/lb | .12 | _____ |
| G = Shearing cost | 2.00 | _____ |

**Table 1.3.   Budget worksheet**

|  | Sample | Yours |
|---|---|---|
| Expenses |  |  |
| Corn $A \times D \times 5$ | $ 500.00 | _____ |
| Hay $A \times E \times 0.4$ ton | 960.00 | _____ |
| Supplement $A \times 20 \times F$ | 96.00 | _____ |
| Total winter feed costs | $1556.00 | _____ |
|  |  |  |
| Shearing $A \times G$ | 80.00 | _____ |
| Veterinary expenses[a] $(A \times B)/30$ | 80.00 | _____ |
| Total expenses | $1716.00 | _____ |
|  |  |  |
| Income |  |  |
| Lamb receipts $A \times 1.2 \times B$ | $2880.00 | _____ |
| Wool[b] $A \times 8 \times C$ | 256.00 | _____ |
| Total income | $3136.00 | _____ |
| Estimate of income from sheep (income minus expenses) | $1420.00 | _____ |

Note: Refer to Table 1.2 for values for A-G.
[a]Estimate based on price of 1 slaughter lamb/30 ewes.
[b]Estimate based on 8 lb wool/ewe.

## Table 1.4. Enterprise budget

| Early lambing (Jan-Feb) | Total | Cash | Late lambing (April-May) | | Total | Cash |
|---|---|---|---|---|---|---|
| **Income** | | | | | | |
| Lambs: 1.25 head × 110 lb × $____/lb | $____ | $____ | 1.35 hd × 110 lb | $____ | | $____ |
| Cull ewes: 0.15 head × 150 lb × $____/lb | $____ | $____ | 0.18 hd × 150 lb | $____ | | $____ |
| Wool: 9 lb × $____lb | $____ | $____ | 11 lb | $____ | | $____ |
| Gross Income | $____ | $____ | | $____ | | $____ |
| | | | | | | |
| **Variable Costs** | | | | | | |
| Feed costs | | | | | | |
| Corn @ $2.00/bu/10 bu | $ 20.00 | $ 0.00 | 8 bu | $ 16.00 | | $ 0.00 |
| Supplement and minerals @ $0.14 × 100 lb | $ 15.00 | $ 15.00 | 60 lbs | $ 9.00 | | $ 9.00 |
| Alfalfa-brome hay @ $65.00/ton × 0.4 ton | $ 26.00 | $ 0.00 | 0.3 tons | $ 19.50 | | $ 0.00 |
| Pasture @ $26.50 per acre × 0.2 acre | $ 5.30 | $ 2.65 | 0.3 acres | $ 7.95 | | $ 3.98 |
| Total Feed Costs | $ 66.30 | $ 17.65 | | $ 52.45 | | $ 12.98 |
| | | | | | | |
| Veterinary, medical, shearing | $ 8.00 | $ 8.00 | | $ 9.00 | | $ 9.00 |
| Machinery and equipment operating | $ 5.00 | $ 5.00 | | $ 4.00 | | $ 4.00 |
| Marketing and miscellaneous | $ 5.00 | $ 5.00 | | $ 5.00 | | $ 5.00 |
| Interest on feed etc., 10% for 6 mo | $ 4.22 | $ 0.00 | 6 mo | $ 3.52 | | $ 0.00 |
| Labor @ $7.00/hr × 6 hr | $ 42.00 | $ 0.00 | 4 hr | $ 28.00 | | $ 0.00 |
| Total Variable Costs | $130.52 | $ 35.65 | | $101.97 | | $ 30.98 |
| Income Over Variable Costs | $____ | $____ | | $____ | | $____ |
| | | | | | | |
| **Fixed Costs** | | | | | | |
| Machinery, equipment, housing, fencing | $ 11.01 | $ 1.10 | | $ 9.79 | | $ .98 |
| Int, ins on breeding flock @ 10% | $ 10.80 | $ 1.08 | 10% annually | $ 10.80 | | $ 1.08 |
| Ram replacement | $ 5.60 | $ 5.60 | | $ 5.60 | | $ 5.60 |
| Total Fixed Costs | $ 27.41 | $ 7.78 | | $ 26.19 | | $ 7.66 |
| | | | | | | |
| Total of All Costs | $157.92 | $ 43.43 | | $128.17 | | $ 38.63 |
| | | | | | | |
| Income Over All Costs | $____ | $____ | | $____ | | $____ |
| | | | | | | |
| Break even prices assume cull ewe income of $5.00 and wool income of $9.00 (early) and $11 (late). Break even selling price for variable costs[a] | $ .85 | | | $ .58 | | |
| Break even selling price for all costs[a] | $ 1.05 | | | $ .76 | | |

Source: Courtesy Iowa State University Extension Service.

Note: These figures assume a 160% lamb crop for early lambing, and 170% for late. A ewe unit is considered to be 1 ewe, .2 replacement ewes, 1.6 lambs, and .04 ram. A ewe flock consists of one ewe unit. Death loss of 10% for lambs weaned and 5% for ewes and ewe lambs is assumed. A ewe flock consists of one live unit.

[a]Assumes cull ewe income of $5.00 per unit and wool income of $9.00 per unit for early and $11.00 for late.

services, people advertising in sheep journals, or neighbors with sheep. Any attempt to give specific figures is bound to be misleading.

Two methods of budgeting, a simple cash-flow or an enterprise budget, may be of interest to the beginner. The cash-flow budget refers to yearly income and yearly expenses. In order to make some estimates, you can use the following process. Start by identifying a possible number of sheep with which to start. You might use the number that you can support on pasture. While this varies, an estimate of four ewes per acre is reasonable. Your extension service, local feed stores, and other sheep producers can help you with estimates for the other prices (Table 1.2.).

The simple budget in Table 1. 3 will help you see how factors such as feed costs and the number of lambs raised per ewe can affect your profits.

The enterprise budget is a way to bring in other information you need to judge the profitability of your sheep operation. Farm businesses use the enterprise budget to compare the worth of different aspects of their operations. In Table 1.4 you will see one kind of enterprise budget that you can use as a worksheet.

The specifics that go into a sheep enterprise budget for an ongoing farm operation will depend on the records being used for other operations. Sample records are given in Chapter 9. However, the most detailed enterprise budget cannot always answer the question of whether sheep will fit into your farm operation. Even a sheep operation that doesn't stack up to other projects on paper can sometimes be justified. For example, sheep may be using forage that would otherwise be wasted.

Profitability may not be the main goal of your sheep operation. In fact, if labor is included in the formula, many sheep producers might smile at the idea of profitability. But even if your operation is a labor of love, you will enhance the "love" part more by using an organized approach to the care and financing of the enterprise.

## REFERENCES

1. Wing, Joseph. *Sheep Farming in America.* New York: Sanders, 1912 p.14.
2. American Sheep Industry Association. *The Sheep Production Handbook.* Englewood, CO: ASIA, 1986.
3. *The Story of Wool.* Denver: The Wool Bureau, 1968.
4. Ensminger, M.E. *Sheep and Wool Science.* Danville, IL: Interstate, 1970.
5. Randall, Henry. *The Practical Shepherd.* Rochester: NY, DDT Moore, 1864.
6. USDA/ Agricultural Statistics Service, 1996.
7. American Sheep Industry Association. U.S. Sheep Industry Market Situation Report. Englewood, CO: ASIA.
8. *Midwest Farm Handbook.* Ames: Iowa State University Press, 1969.

# 2. Buying Sheep

*"Buy disease-free animals of a breed or breed mix adapted to the climate."*—STAN AND JEAN POTRATZ

*"We bought everyone else's problems."*
—JAN AND BILL BUTLER

As a beginner, buying sheep will be a Catch 22. To make a good purchase, you need the experience and knowledge you will get only after you buy sheep and work with them. What breed is best for you? How can you tell a healthy sheep from an unhealthy one? What does a good sheep look like? Who should you buy from? When should you buy? How many should you buy?

From Chapter 1, you should know a little bit about the costs of a sheep operation, the possible income, and a bit about labor requirements. You may be thinking about starting with just a few sheep. If you have the facilities and pasture (see Chapters 3 and 4 for requirements), a larger number will not increase your workload a great deal and will provide more return and more experience for your investment. A flock of 40, or what is called a one-ram flock, is a manageable enterprise for a serious beginner.

## KINDS OF SHEEP

Breeding, conformation, and age are all important in choosing your sheep. No one set of terms will describe these features. Make sure what definition is being used when you go to buy. The terms *blackface, white-*

*face, western, native, purebred, crossbred, and grade* are common terms in addition to the names of specific breeds.

## Blackface

Blackface sheep have black, black and white speckled, or brown faces. The blackface, which usually has some Hampshire or Suffolk breeding, was bred for meat rather than wool production and includes the larger breeds.

## Whiteface

Whiteface sheep probably have some combination of Corriedale, Columbia, or Rambouillet, or some of the smaller whiteface breeds such as Finnsheep and Cheviot. The whiteface sheep tend to be smaller than the blackface sheep and produce more wool. These black- and whiteface characteristics will vary considerably with the exact breeding.

## Western

Westerns are sheep that come from ranches in the West as replacement ewes. Western sheep can be either blackface or whiteface, but the whiteface, heavier wool-producing breeds are most commonly known as westerns.

## Native

Native sheep are raised locally in the farm-flock states. Since the farm-flock states concentrate on meat production, these sheep are most apt to be blackfaces. Technically, native and western refer to where the sheep are raised and blackface and whiteface refer to how they are bred. In practice, however, blackface sheep are often referred to as natives and white-face sheep as westerns, regardless of where they come from.

## Grade

Grade sheep come from a registered purebred ram and a ewe that looks like the same breed as the ram but is not a purebred. Some sellers use the term *grade* to refer to a group of ewes that resemble a certain breed, regardless of whether the ram was registered or not.

## Crossbred

Crossbred sheep come from parents of two or more different breeds. Some desirable crosses will be discussed in Chapter 4. Sometimes sellers loosely use the term *crossbred* to mean any sheep that are not purebred. The crossbred offspring from purebred parents with records is, of course, the more valuable animal.

## MAKING YOUR SELECTION

Your choice of sheep can make management easier or harder. Whenever possible, buy ewes with similar breeding. You want breeding times, lambing seasons, and growth rates in your flock to match as closely as possible. Select ewes of a uniform size; feeding will be more efficient if they require similar amounts of feed to stay in good condition.

If ewes are of similar genetic background and age, they can be bred at about the same time. This will keep your lambing seasons shorter. Lambs will probably have similar growth rates and will reach market weight at about the same time. The more uniformity the easier the management.

The type of sheep you chose will depend on the goals of your operation and what products you are interested in selling. Generally, you will want to select breeding stock for both good wool and meat production. One such combination is to cross a whiteface ewe, with good wool production, heartiness, and medium size, with a blackface ram, known for its good meat production. Grade or crossbred ewes of uniform size and appearance seem to be the best choice for the beginner.

### Selection by Age

Terms referring to the sheep's age include *ewe lambs, ram lambs,* and *yearlings* or sometimes just *mature.* Ideally, the beginner should buy yearling or 2-year old ewes. They have their first, often frustrating, lambing experience behind them, and they have long reproductive lives ahead of them. They will, however, be the most expensive to purchase.

Several factors should be considered in selecting ewes by age. Ewe lambs, ewes that have not been through a breeding season and have their whole reproductive lives ahead of them, will be more expensive than 4- to 6-year old ewes. They can be among the most difficult to handle during their first lambing. First-time mothers can exasperate even experienced shepherds by gazing at a struggling newborn lamb in disbelief and then walking off to feed. Extra guidance is sometimes needed to get the new mother started properly. The rate of multiple births is lower in the first lambing, and a small number of perfectly sound ewe lambs may miss the first breeding season altogether.

Older ewes have their own problems, and great care should be exercised when purchasing them. Although a rare ewe may be productive into her teens, most sheep producers cull heavily among the 7- to 8-year olds. Culls are ewes being sold because they are barren, have bad udders, or require extra care. Older ewes tend to lose their teeth; consequently, many need special care in feeding.

Until about age 4, the sheep's age can be determined by the number and kind of front teeth. The 4-year old will have eight permanent incisors. After age 4, the sheep may keep a full set of teeth for some time, or the teeth may wear down and spread. Some may be lost. Old sheep with no teeth at all are called gummers. Try to get some practice in judging the age of sheep from the teeth. Figure 2.1 will serve as a guide.

A critical factor in determining what you should pay for a ewe is the probable number of lamb crops she will produce. A healthy ewe lamb will produce seven lamb crops. Most older ewes will cost less, and will produce fewer crops. The extra cost of yearling or 2-year-old ewes is usually worth your investment.

Conformation features are important in choosing sheep because good features do not result simply in a good looking sheep. They are related to reproductive abilities and meat production. Look at the height, length, and depth of the body. The sheep should be well proportioned. Each breed has its own conformation standards, but Figures 2.2–2.6 illustrate general features to look for. These features apply to both ewes and rams, although the ewe will have more delicate features.

The breeding and nursing ability of the ewe is not always identifiable. A ewe may have excellent conformation and be in good condition and still be barren. Younger ewes in a sale barn should raise the question of whether they are being sold because they weren't productive breeders. Always examine the condition of the ewe's udder for signs that she is unable to nurse. This is another reason why a healthy-looking young ewe

**2.1.** Frontal views of the lower jaws of sheep of various ages. The upper jaw has no incisor teeth in the front. **A.** All of the lamb's teeth are small. These temporary teeth are shed to make way for permanent teeth. **B.** Mouth characteristics of yearlings; the two large teeth in the center are permanent incisors, which come in immediately following loss of temporary central pairs. **C.** Two pairs of permanent incisors, characteristic of 2-year olds. **D.** Mouth of 3 year old, three pairs of permanent incisors. **E.** Full mouth characteristic of a 4-year-old.

At 5 or 6 years of age, these permanent teeth wear down and appear more slender with flatter grinding surfaces. At variable ages, depending on breed, teeth begin to spread and loosen through the "spreader," "broken mouth," and "gummer" stages. Sheep have no teeth at all at the gummer stage. Large growthy animals often replace temporary incisors with permanent incisors earlier than the ages given above. *(Courtesy, American Sheep Industry Association)*

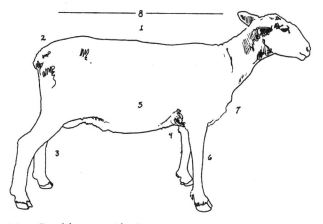

**2.2.** Good features, side view.
   1—Straight topline
   2—Some downward slope of rump for easier lambing
   3—Good substance of bone, hock not too straight
   4—Good girth in heart area
   5—Good girth through middle
   6—Front legs straight
   7—Trim brisket
   8—Good length of body

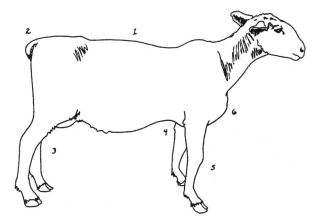

**2.3.** Features to avoid, side view.
   1—Topline rough
   2—Rump too level; may contribute to lambing problems
   3—Hock too straight
   4—Not enough depth in forerib
   5—Front legs crooked, weak pasterns
   6—Brisket too prominent

**2.4.** Good features, top view.
   1—Shoulders trim
   2—Good width across rib cage and loin
   3—Slightly narrower across shoulders than across rump
   4—Good width in rump
   5—Straight legs

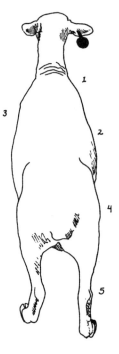

**2.5.** Features to avoid, top view.
   1—Shoulders coarse, sticking out
   2—Little width across rib cage
   3—Narrow rump
   4—Crooked legs (sickle and cow hocked)

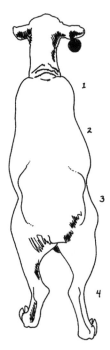

**2.6.** Bad and good features, front and hind view.
1—Chest too narrow, legs too close together
2—Shoulders coarse, sticking out
3—Good width of chest, good set to legs
4—Trim shoulders
5—Good straight legs, good-sized legs
6—Good width in pelvic area
7—Narrow in pelvic area
8—Hocks set in

might be culled. A healthy udder of a ewe that is not nursing will be soft, pliable, and symmetrical. Two normal-sized teats are needed to support two nursing lambs. The novice buyer should look for evidence of mastitis, a common infection that occurs during lactation. Mastitis can cause permanent lumps and unevenness in the udder and make the ewe unable to nurse a lamb. Figures 2.7 and 2.8 provide examples of sound and unsound udders, respectively. Ewes may also become unable to nurse lambs if they have lost a teat due to careless shearing. Ewes without two teats or with unsound udders are not good buys.

## Selection of Rams

The best ram you can afford should be used in the breeding of commercial ewes. A large, meaty, blackface ram is a good choice. The sire can strongly influence the lambs' ability to gain weight. A ram with a record of weight gain of l pound a day—120 pounds in 120 days—is considered

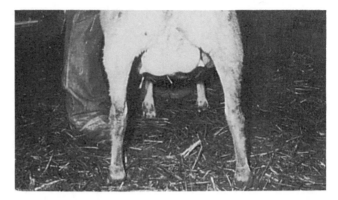

**2.7.**    Sound udder.

a good standard. The best way to be sure you are buying a ram with this trait is to buy one with records, and check the records. Breeders of purebreds are the most likely to keep these records. If records are not available for a ram, look for a large, well-muscled animal that is free from conformation defects; the ram should weigh 250 pounds or more at a year. The best time to judge a ram is to view him after shearing.

Obviously, you want a fertile ram. Look for normal-sized testicles, using other rams for comparison. Many breeders of purebreds give some kind of assurances that the ram is fertile. It is certainly appropriate to ask

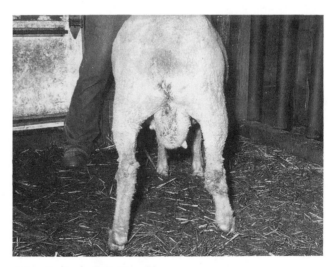

**2.8.**    Faulty, undesirable udder.

about fertility. Ask, for example, if the ram has been overheated, one cause of infertility. Whenever records are available, check them for any history of such problems as rectal prolapses, inverted eyelids, and unsound mouths. Although the particular ram you are interested in may appear sound, evidence of these problems may be in the bloodline. Epididymitis is a disease that can also cause infertility and rams can be tested for it. See Chapter 6 for a description.

It may not be economically feasible to buy a good registered ram to service a few ewes. Rather than settle for a cheap, inferior ram, consider alternatives: have your ewes serviced elsewhere and pay a stud service for a good ram, or share in the purchase price of a good ram with another small-flock owner.

## HEALTH PROBLEMS

Before you buy your first sheep, learn to identify the most troublesome health conditions. A tutor in the form of a more experienced shepherd is of special value at this time.

Topping the list of health problems is footrot. This is a highly infectious disease that begins in the soft tissue of the hoof. The hoof gets soft and oozy and separates from the underlying soft tissue. Footrot can also be identified by its offensive odor. The disease causes serious lameness, and the infected sheep has a difficult time feeding and keeping up with the flock. As a result, it loses weight.

When buying sheep, you should examine the hooves thoroughly. If you are buying so many sheep that you cannot examine each one individually, look for any sheep that are limping and examine them. But be aware that in the early stages, footrot does not cause lameness. If you discover one case of footrot, be suspicious. The disease is so contagious that the whole flock may be infected. It is possible to treat footrot (discussed in Chapter 6), but it is difficult, messy, and time-consuming, and sadly, the animals are in pain.

You should also watch for parasites in your prospective sheep. Extremely thin and listless sheep or those with runny noses and coughs may have some kind of internal parasite. External parasites, such as ticks, can be seen with the naked eye. Look for them by parting the wool around the neck. If wool is falling off or if the sheep is rubbing itself, it probably has some kind of external parasite. While shepherds expect to treat sheep for parasites, a sheep that is seriously infested is hard to get back in good health. Some can even die before the treatment takes effect.

These health problems are covered in more detail in Chapter 6.

## THE TIME TO BUY

Typically, ewes are purchased from midsummer to early fall to allow them time to adjust to their new home and to prepare them for the breeding season. Sometimes spring or midsummer markets provide a better selection at lower prices. Some pregnant ewes may be available in the late fall, but the selection is small and the prices are higher than those for unbred ewes.

If special opportunities arise—say, an elderly couple decides to go to Florida and wants to sell their sheep quickly—be ready to act. At the same time, don't allow yourself to be so carried away by a special opportunity that you settle for inferior stock.

## THE PLACE TO BUY

Regardless of where you look for sheep, you are going to be uneasy about paying too much or buying low-quality stock—or both. The more you learn about sheep, the less you will have to rely on your intuitions about the honesty of the seller. It takes some effort and time to learn about the signs of age, health, and productiveness of the animals. In the beginning, going through established channels rather than looking for suspiciously good buys is the best bet. Here are some general buying tips:

1. Tune into the grapevine through local shepherds. While you want to avoid unfair gossip, you can learn much from reputable producers. A dealer who has sold culls to unsuspecting buyers soon gains a reputation for dishonesty. Established sheep producers have an interest in maintaining the reputation of the sheep industry in general. Through your extension service, sheep producers' associations, or neighbors, find knowledgeable people who have no vested interest in your sheep business and check out buying possibilities with them.

2. Be wary of sale barns. The sale barn is traditionally the place to sell culls. If you have reliable information that a good flock is going to be sold, you may be on solid ground, but you need to know more about the sheep than what you see in a sale ring.

3. Check with your extension service or local sheep organizations about special opportunities to buy sheep. In some areas, sheep producers get together every fall to buy semitrailer loads of young western ewes for replacements. They have an established source, and often the ewes will be vaccinated and wormed on their arrival. These kinds of arrangements allow the beginner to buy along with more-experienced sheep pro-

ducers. Some states have replacement-ewe directories where people can list sheep they have for sale. Your reputable sheep advisor can look at the directory with you and point out good producers, as well as farms to avoid.

4. Don't rush through an examination of sheep because the seller seems impatient or annoyed with your fussiness. The reputable seller will understand and respect your concerns. Don't be intimated because you are a beginner; this can only benefit the person selling poor-quality sheep.

5. Ask to see records on the sheep. Many flock owners do not keep records, but it is a good sign if they do. Simple records noting birth weight, whether they were twin or single births, and rate of growth can help you select quality sheep. Records also indicate that ownership of the sheep has been taken seriously and that there has been a general policy of good care.

6. Insist on getting what you ordered. If you placed an order specifying age, breed, and quality and the ewes aren't what you expected, don't take them.

7. Try to establish the sheep's age. This is how you estimate the value of the ewe to your operation—that is, how many lamb crops you might reasonably expect.

8. Reject droopy, dejected, extremely thin ewes and those that are not breathing normally.

## THE PRICE TO PAY

Local supply and demand affect the cost of your sheep. Breeding and age are also important factors. One rule of thumb is that the price of the ewe should equal what she will gross in 1 year in lambs and wool. Another formula is to pay 1.25 times the cost of a market lamb. These formulas may work in some areas and not in others. Learn about the price of sheep in your area, as well as the price of western ewes. Relate your information about price to the age and type of sheep being sold. Some additional value can be assigned to animals known to be twins or fast growers. Don't judge the worth of the sheep just by the appearance.

Give yourself time to learn what you need to know, to familiarize yourself with local opportunities, and to get a firsthand look at some health and breeding defects before you try to make a purchase for yourself. Although your eye will improve with experience, even as a beginner, you can avoid the important health and breeding problems. The care you take in selecting sheep will pay off as your shepherding continues.

# 3. Shelter and Equipment

*"Feeders need to keep the shepherds from the sheep. Walking through a flock with a bucket can be dangerous."*—STAN AND JEAN POTRATZ

*"The first year, we didn't have corrals and we just lambed out. It didn't work out as well."*
—DEBBIE BROWN

The longer you work with your sheep, the more you understand what the ideal arrangement of equipment and shelter should be. Your equipment need not be new or mechanized. You should locate grain bunks close to feed storage areas. Sheds and barns should be equipped with electricity and running water and easy to clean. You want tight fences. And, you need a place where sheep can be corralled easily and moved to clear pasture without having to herd them—you don't want to go through five gates and your neighbor's flower bed. You need ways to safely handle sheep during the administration of medications as well as shearing so neither you nor the sheep are injured.

While sheep producers can move closer to their ideal setup every year, few have been able to justify starting out with textbook-efficient buildings and equipment. Making-do and improvising are traditions with sheep raisers. Because of their hardiness and adaptability, sheep can be raised successfully with a minimal amount of special equipment and housing. The fact that some range flocks are raised with little shelter should give hope to the person who thinks it is not possible to raise sheep without proper facilities.

No matter what size flock you are starting with, planning is of the utmost importance. Providing adequate fencing, water, feed, shelter, and work areas will spare you last-minute problems during lambing and will prevent that crew of impatient shearers from having to wait while you build a corral for your sheep. This chapter discusses the basic housing and equipment requirements needed for a successful operation. A guide for planning your own sheep facilities will also be provided.

## FENCES

Sheep are easier to keep fenced than are cattle or hogs. They are peaceful creatures that like to establish their own patterns of grazing, unless they are not being fed enough. It pays to make fences secure before your sheep start investigating and testing them. Permanent and flexible fencing can be combined on your operation.

Electric fences can be used with sheep and are especially helpful in setting up grazing areas where you do not want a permanent fence. The advent of flexible net electric fencing that is easy to move and set up has made it possible to more fully take advantage of the sheep's special foraging abilities.

Because wool insulates the sheep from electric shock, sheep should be exposed to electric fence shortly after they are sheared. This way they will learn to avoid the fence before their wool grows back and protects them. The training may need to be repeated on a yearly basis.

The training may take place by putting food just below and beyond the wire or even by tying ears of corn to the fence with baling wire. Some people recommend attracting the sheep to the wire with something shiny, such as aluminum pie pans. But remember, electric fences can be very dangerous, even deadly. Buy only approved equipment and follow carefully the directions for installation.

If you are building a new fence, standard recommendations for fence posts and wire can be obtained from your local extension office. Since buying the best and most long-lasting fence is expensive, you may have to do some compromising. Use your best fencing around corrals or feed areas where sheep are more likely to rub and push. Unfortunately, you may not know where the areas of stress are until you have been through a season with sheep that lean against hastily erected or mended fences.

Divide your pasture for the most efficient use of your space. In your first year, experiment with temporary and/or inexpensive fencing arrangements until you have an idea of your pasture rotation needs. Several types of portable electric net fencing designed for sheep (Fig. 3.1) can be or-

**3.1.**   Portable electric fencing.

dered from the catalogs listed in Appendix B. A year or two of experience will give you a better idea of how many sheep your pasture can support, where you want to put permanent fencing, and how you can coordinate your fencing needs with your existing buildings.

## SHELTER

If you have any farm buildings on your acreage, you can probably adapt them for sheep. Open-front machine sheds, big barns, large chicken houses, small hog buildings, or part of a big metal machine shed are all adaptable. Consider your buildings according to the major shelter needs of your sheep—winter shelter for the ewes, a lambing or maternity area, and shelter for weaned lambs on feed. These needs do not occur simultaneously. For example, by the time your lambs are weaned, you may be able to put them in your winter ewe shelter, while the ewes go out to pasture. Therefore, consider which buildings can be used during which parts of the year. Is there a farrowing house that will be empty during lambing? If you have cattle, will your bottle-fed calves be out of their stalls in time to use that area for feeding young lambs?

Sheep are naturally equipped to endure cold, thus, they are generally sheltered in cold housing except on the rare occasions when new lambs require a source of heat, such as a heat lamp. A more serious concern is the need to provide proper ventilation in sheep buildings. One hundred sheep give off about 20 gallons of water a day; in a poorly ventilated shelter, this water loss poses serious health hazards and contributes to the threat of pneumonia. Open-front sheds are preferable to tightly enclosed buildings. If you must use closed barns, keep the doors open much of the time to ensure good ventilation.

Providing proper shelter for all your sheep during lambing can be a problem. Pregnant ewes should not be closely confined inside, because they need exercise. At the same time, it can be a problem when lambs are born outside in below-zero temperatures. As a rule, the ewe will seek protection when she needs it, and methods for getting her inside before she actually lambs are outlined in Chapter 7. Ideally, the mother should have exercise and plenty of room until she delivers, give birth inside a shelter, and be penned up with her lamb right after delivery. For the best survival rate, the mother and new lambs should be put inside a 4-by-4-foot or 4-by-6-foot pen called a lambing jug; this jug provides protection from the cold and ensures that the mother and lambs will form a bond. Ewes that are about to lamb can be moved inside, while the rest of the pregnant ewes should be allowed more freedom in sheds and lots.

When confined alone in a shelter, a ewe will need 16–20 square feet or 10–12 square feet in an open-front building, as well as 25–40 square feet of lot space. A ewe with lambs in confinement will need 18–20 square feet or 12–16 square feet in an open-front building, as well as 25–40 square feet of lot space. Feeder lambs need less space, depending on their size.

After the ewe is confined in a pen with her lambs for a few days, she needs to be gradually introduced into larger groups of mothers and lambs. A flexible arrangement of wooden panels or gates can help you mix new mothers and lambs, first in groups of two or three ewes, then in groups of four to six. The more you can control bonding of mothers and lambs, the less likely it is that the lambs will be abandoned.

Whatever buildings you decide to adapt for your sheep, first clean them out thoroughly and make sure there are no mysterious cans of paint or chemicals present. Rid the building of bailing string and wire and broken tools that can be hazards to both the sheep and the shepherd. You may want to use a disinfectant, depending on the building's previous use. Check with your local extension service or veterinarian to learn about potential health hazards.

## FEED AND WATER

Provide clean feed and water for sheep in a manner that keeps feed off the ground and away from feces and that prevents the sheep from rubbing wool off their necks and imbedding feed in their wool. Ewes need 16–20 inches of feeder space each; you can water 15–25 ewes for every foot of the perimeter of a tank. Automatic waterers can serve as many as 50 ewes and lambs per bowl. Ideally, running water should be available in the buildings you use, but when you need to haul water, you can delay freezing by packing the water tanks in straw or dirt or by using a small stock-tank heater.

In addition to water and feed, you will need to provide a feeder for salt and minerals. Designs for basic grain and hay feeders and a salt feeder are shown in Figures 3.2–3.4, though you can improvise, depending on the lumber you have available and the location of the feeders. Keep your eye out for good ideas when you visit other farms. Remember to keep your feeders sturdy because they will have to withstand a considerable amount of rubbing and pushing.

Lambs will need slightly modified equipment (Figs. 3.5 and 3.6). Many sheep producers feed their lambs from creeps at an early age. A creep is an area that lets the lambs in but keeps the ewes out. In a creep, lambs have access to feed all the time. You can set up a creep area either inside or outside, depending on the weather and your facilities. The essential ingredient for a creep is a panel with adjustable vertical slats spaced from 6 to 9 inches apart depending on the size of the lambs.

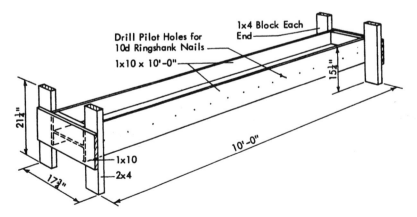

**3.2.** Grain feeder for ewes. (*Courtesy, MidWest Plan Service*)

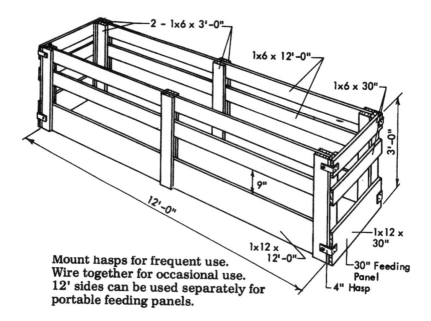

2 – 1x6 x 3'-0"

1x6 x 12'-0"

1x6 x 30"

3'-0"

9"

12'-0"

1x12 x 12'-0"

1x12 x 30"

30" Feeding Panel

4" Hasp

**Mount hasps for frequent use. Wire together for occasional use. 12' sides can be used separately for portable feeding panels.**

**3.3.** Hay feeder. (*Courtesy, MidWest Plan Service*)

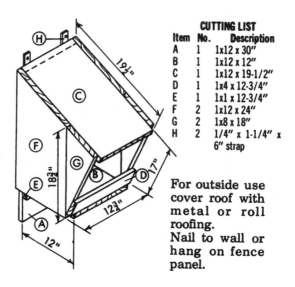

**CUTTING LIST**

| Item | No. | Description |
|------|-----|-------------|
| A | 1 | 1x12 x 30" |
| B | 1 | 1x12 x 12" |
| C | 1 | 1x12 x 19-1/2" |
| D | 1 | 1x4 x 12-3/4" |
| E | 1 | 1x1 x 12-3/4" |
| F | 2 | 1x12 x 24" |
| G | 2 | 1x8 x 18" |
| H | 2 | 1/4" x 1-1/4" x 6" strap |

19½"

18¾"

17"

12¾"

12"

**For outside use cover roof with metal or roll roofing. Nail to wall or hang on fence panel.**

**3.4.** Salt and mineral box. (*Courtesy, MidWest Plan Service*)

34

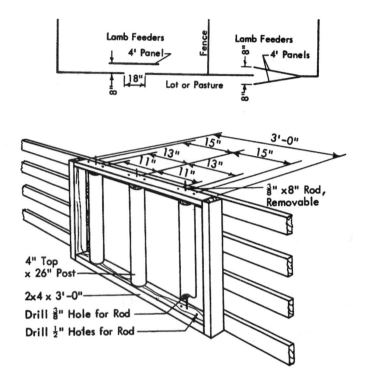

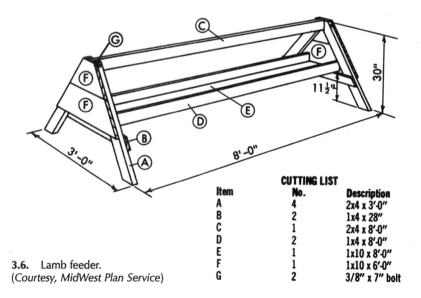

**3.5.** Lamb creep, using lambing pen panels. (*Courtesy, MidWest Plan Service*)

**3.6.** Lamb feeder.
(*Courtesy, MidWest Plan Service*)

**CUTTING LIST**

| Item | No. | Description |
|------|-----|-------------|
| A | 4 | 2x4 x 3'-0" |
| B | 2 | 1x4 x 28" |
| C | 1 | 2x4 x 8'-0" |
| D | 2 | 1x4 x 8'-0" |
| E | 1 | 1x10 x 8'-0" |
| F | 1 | 1x10 x 6'-0" |
| G | 2 | 3/8" x 7" bolt |

## PANELS AND CORRALS

The shepherd will need a number of multiple-use panels of different sizes. To make lambing jugs, 4-by-4-foot, 4-by-5-foot, or 5-by-5-foot panels are needed, depending upon the size of the ewes. You should have enough panels to make pens for 10–15% of your ewes. Figure 3.7 (facing page) shows two ways to use the panels. Some people prefer solid panels because they help keep drafts off the baby lambs. You may find panels easier to move if they are not permanently hinged together. The panels can be used later in the construction of creep areas.

It is also helpful to have a number of longer panels, 8, 10, or 12 feet in length and relatively light in weight. These panels can be used for separating mothers with twins from mothers with singles, setting up space for shearing, mixing the mothers and sorting lambs, or doing routine health maintenance chores.

## HEALTH CARE AND MISCELLANEOUS EQUIPMENT

Whether you have a few sheep or a large flock, your routine health care needs will be the same. If you have a few sheep, you can get by with very simple equipment, but if you have a large flock of sheep, you may want to invest in equipment that will get the jobs done more quickly and efficiently. Suppliers of sheep and veterinary equipment are listed in Appendix B; you can send for their catalogs. Be prepared to invest in the following:

1. Hoof trimmers. Hoof trimmers (Fig. 3.8) are essential for any sheep owner. Trimmers can be obtained from most veterinary supply houses. Hooves must be trimmed at least yearly and sometimes more often.

**3.8.**  Hoof trimmers.
(*Courtesy, Mid-States Livestock Supplies*)

2. Flashlight. A good flashlight is very important, especially during lambing. Trying to get by with a two-battery pocket flashlight won't do, especially if any of your buildings are without electricity. You need a flashlight that can give you a good look at your corrals and buildings without making you trudge over every inch on a dark night.

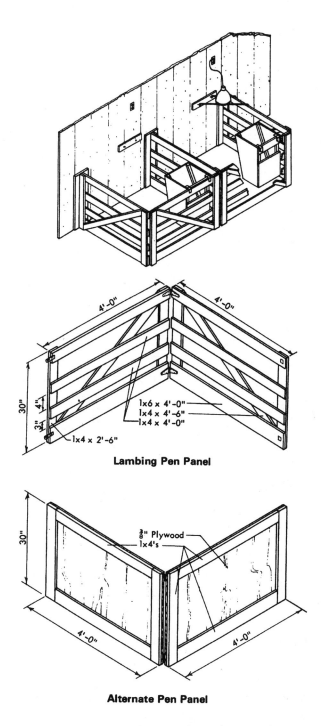

1x6 x 4'-0"
1x4 x 4'-6"
1x4 x 4'-0"

1x4 x 2'-6"

30"

4"

3"

4'-0"

4'-0"

**Lambing Pen Panel**

$\frac{3}{8}$" Plywood
1x4's

30"

4'-0"

4'-0"

**Alternate Pen Panel**

**3.7.** Lambing panels. (*Courtesy, MidWest Plan Service*)

3. Equipment for worming. Chapter 7 discusses the advantages and disadvantages of different methods of worming, a procedure that must be performed several times a year. You will need equipment for one or two of these methods. All the necessary equipment can be purchased from the catalogs listed in Appendix B.

4. Drenching equipment. Drenching equipment, for delivering liquid medication orally, can range from inexpensive single-dose devices (Fig. 3.9) to ones with portable tanks and that dispense automatic doses (Fig. 3.10).

5. Balling gun. A balling gun is for oral delivery of pills (Fig. 3.11).

6. Injection equipment. Injection equipment is needed for routine treatment of tetanus, overeating disease, worms, or other illnesses. Again, the equipment can range from inexpensive disposable syringes to more-expensive devices that hold large quantities of vaccine and dispense the right dose automatically.

7. Docking and castration equipment. You must first decide on a method for docking and castrating your sheep: elastrators that stretch

**3.9.** Single-dose drenching device. (*Courtesy, Mid-States Livestock Supplies*)

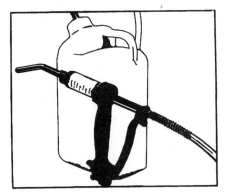

**3.10.** Automatic drenching device. (*Courtesy, Mid-States Livestock Supplies*)

**3.11.** Balling gun. (*Courtesy, Mid-States Livestock Supplies*)

small rubber bands so they can be placed around the tail or scrotum, hot chisels, emasculators, or other methods (see Chapter 7).

8. Means of identification. Plastic ear tags with numbers that can be seen from a distance and are easy to insert can be used to identify your sheep. If you are going to apply the tags, you must buy the insertion device of a particular make when you buy the first batch of numbers. Small flat metal tags (Fig. 3.12) are also used, especially with purebred sheep where detailed records must be kept.

**3.12.**  Metal ear tags.
(*Courtesy, Mid-States Livestock Supplies*)

To make sure there are no mix-ups, the lamb and its mother can be marked with lanolin-based wool paint (which won't hurt the wool) as soon as the baby lamb is dry. This paint is available from supply catalogs. Livestock-marking crayons of bright colors are used to temporarily identify sheep that have been treated for a medical problem.

9. Shearing tool. A small hand-shearing tool can be quite useful even though you may not want to learn to do the actual shearing yourself. Because the shearer's schedule is often full and to save time, if you have a small flock, you may at least take on the job of tagging your ewes (shearing around the vulva and udder) before breeding and lambing. Also, if you are not going to tan your own hides, you may want to shear the lambs you are butchering for yourself before slaughter. This keeps the wool from getting in the way during skinning.

10. Medicine chest. You will want to have a sturdy chest in which to carry your foot trimmers, iodine spray, pine tar, lambing supplies, and other basics.

11. Sheepdogs. Many experienced shepherds consider the sheepdog the piece of "equipment" they would least want to give up. Training a puppy to be a useful helpmate requires patience and time; but once trained, the sheepdog allows you to handle many more sheep than would be possible on your own or even with a number of human helpers. If you feel you don't want to invest the time or money in a sheepdog, make a point of seeing a good dog work with sheep at a fair or on someone else's farm. It is an impressive sight.

As you develop an idea of what you need to make your operation run smoothly, you may want to purchase additional pieces of equipment, such as a loading chute, a foot bath, or a ewe stanchion. Designs for building these and other pieces of equipment can be found in *The Sheep Housing and Equipment Handbook*, listed at the end of the chapter.

## LAYOUT OF YOUR SHEEP AREA

Modest equipment and good planning will work better than expensive equipment and poor planning. If you are a beginner, try to hold off making costly decisions about sheep housing and equipment until you better understand what is involved. Your idea of what is convenient may change a great deal in a year. As you make decisions about how to use buildings or where to spend limited funds, remember that your own preferences are important. You may prefer hauling water to an open shed rather than shovelling manure out of a building that has water but is inaccessible to your loader tractor. A written plan for your first year will be helpful. Figure 3.13 provides a model for your layout. Make rough drawings of your buildings and think about the following questions:

1. What is the square footage of building space you have available in each building? (Write the measurement on your drawing).
2. Given the number of ewes you have decided on, where will there be space to give the ewes winter shelter? (You will find the winter shelter space requirement for ewes on page 32.)
3. How many lambing pens will you need? Where is the best place to set them up? (Sketch them on your plan.)
4. Where will you move the ewes and lambs after they get out of the lambing pens? Where will you set up a creep feeding area while the lambs are still nursing? (Sketch a creep area into your plans.)
5. Where will your waterers be?
6. What length of feed and hay bunks will you need, and where will you put them? (Draw them into your plan.)
7. Where will your feed be stored? Is it located fairly close to your feeders?
8. How will you remove manure from the building?
9. Is there a corral area for shearing your sheep, docking, and other health work? (Draw in the permanent fences you have around your buildings now.)
10. Can you see areas where new fences or temporary panels will be

**3.13.** Model layout for sheep area (facing page).

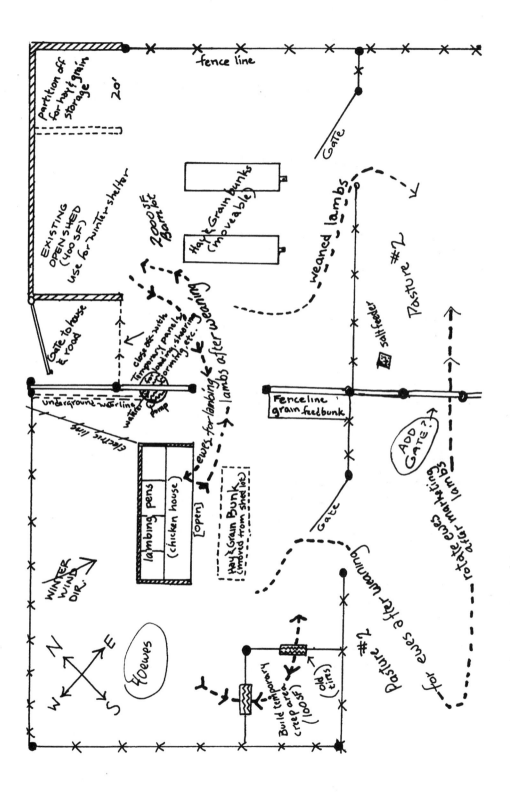

needed? (With a dotted line, trace how you will move ewes from pasture to winter shelter and from the lambing area to lamb feedlots.)

Take time to analyze your situation and begin with the basics. Work out a plan, then ask more experienced shepherds to take a look at it and advise you on simple improvements. Convenience does not always involve spending money—it may be just a matter of creative planning. Keep your plans flexible.

## REFERENCE

*Sheep Housing and Equipment Handbook.* Ames: Iowa State University, Mid-West Plan Service MWPS-3, 1994.

# 4. Breeds and Breeding

*"I don't believe any breed of livestock can be viable unless there is a commercial base."*—Mary Ann Nipp

*"The sheep may look pretty, but is it productive?"*
—Tom Clayman

Beginners may think a study of sheep genetics is too technical, or perhaps even a bit pretentious, for them to consider in planning their first flock. But even the small-flock owner will soon wonder why some apparently similar lambs on the same feed gain weight rapidly while others do not, or why certain ewes seem to have twins regularly and others do not. Even if your understanding of those strange-looking strings of genes may be incomplete, the science of genetics can help you improve your flock. Selective breeding, especially for traits that affect the profitability of the flock, should not be the exclusive concern of breeders of purebreds or managers of experimental flocks.

While the genetic goals of your particular flock may vary, it is useful to think in terms of ewe and ram breeds for your breeding program. Ewe breeds should excel in traits related to reproduction and the ability to mother, whereas the ram breeds should excel in traits related to rate of gain and carcass quality. For commercial production, ewes should be crosses of the ewe breeds. The majority of these are mated to rams of the ram breeds to produce market lambs, and the minority are bred to rams of the ewe breeds to produce replacement ewes. Over 30 breeds of sheep are raised in the United States. This chapter covers the 11 breeds that most typify the ewe and ram traits.

## FAVORED EWE BREEDS

The principal ewe breeds are the Columbia, Dorset, Finnsheep, Rambouillet, Romanov, and Targhee. A good breeding program does not need purebred ewes, but genetic influence from these breeds will improve traits important to the ewe. Western ewes, for example, often are used in commercial production because they are crosses of two or more of these ewe breeds.

### Columbia

The Columbia sheep (Fig. 4.1) is an American original developed in Idaho in the early twentieth century by breeding Rambouillet ewes with Lincoln rams. The Columbia was developed for range conditions (hardiness and strong flocking instincts), but it is used successfully in farm flocks also. It has good wool production, and the mothers are fairly prolific and are good milkers. Selection also has stressed their large mature size, so they also can be used as a ram breed to produce big market lambs.

### Dorset

The Dorset (Fig. 4.2.) is a medium-wool breed that comes originally from England. It has become more popular since a hornless strain was developed. Dorsets are medium-sized sheep, and the ewes are excellent milk producers. They have a long breeding season and have a tendency for fall lambing, for people interested in accelerated lambing (lambing more often than once a year). Another attractive characteristic of the Dorset is that the face, ears, and legs are practically free of wool.

### Finnsheep (Finnish Landrace)

Finnsheep (Fig. 4.3) were imported into Canada in 1966 and into the United States in 1968. Even though this breed does not rate high in either growth or wool production, the commercial breeder is attracted by its high lambing rate: 250–450% lamb crops are common (a 200% lamb crop means ewes average two lambs). Many commercial sheep producers introduce some Finnsheep blood into their flocks to improve the lambing average. Estimates show that the percentage of Finnsheep in a crossbred ewe will result in an equal percentage of increase in the lamb crop.

### Rambouillet

The Rambouillet (Fig. 4.4) is a breed that was developed from the Merino by the French to improve the carcass quality while maintaining

**4.1.** Columbia. (*Courtesy, American Sheep Industry Association*)

**4.2.** Dorset, polled and horned. (*Courtesy, American Sheep Industry Association*)

**4.3.** Finnsheep. (*Courtesy, American Sheep Industry Association*)

the good wool and mothering features of the Merino. The Rambouillet is often the foundation stock for range flocks in the United States because of its strong flocking instincts and hardiness. It also has an extended breeding season that starts in July or August and continues into April or even May.

## Romanov

The Romanov (Fig. 4.5) originated in Russia and was imported into Canada in 1980 and from there to the United States in 1986. They have a gray to black fleece with a high percentage of hair fibers. They are closely related to the Finnsheep breed and, like the Finnsheep, have a high lambing rate. Research data suggest that the lambing rate of the Romanov is even higher than that of the Finnsheep.

## Targhee

The Targhee (Fig. 4.6) is another popular breed of range sheep that

**4.4.** Rambouillet. (*Courtesy, American Sheep Industry Association*)

**4.5.** Romanov. (*Courtesy, American Sheep Industry Association*)

was developed primarily from the Rambouillet and has much of the Rambouillet's adaptability to range conditions. Targhees were bred for long wool, increased lambing percentage, and milk production.

## FAVORED RAM BREEDS

The ram breeds, emphasizing rate of growth and carcass quality, mostly originated in England and include the Suffolk, Hampshire, Oxford, and Texel.

### Suffolk

The Suffolk (Fig. 4.7) was developed in England and imported to the United States in 1888. It is one of the most distinctive looking breeds, with its wool-free black face and long ears. The fleece is perhaps the Suffolk's weakest point, for it is fairly light and has a tendency to have black fibers. The strong points of the breed are the growth rate of the lambs, its good

**4.6.**  Targhee. (*Courtesy, American Sheep Industry Association*)

**4.7.**  Suffolk. (*Courtesy, American Sheep Industry Association*)

carcass characteristics, and its aggressive breeding behavior. The Suffolk ewe is prolific and a good milker.

## Hampshire

The Hampshire (Fig. 4.8) and the Suffolk are the best-known meat-producing breeds. Hampshire fleeces are a bit better than those of the Suffolk in both quality and quantity, but they also have some black fiber. Hampshire lambs are well known for their rate of growth and early finishing characteristics.

## Oxford

The Oxford sheep (Fig. 4.9) produces a heavier fleece than either the Hampshire or Suffolk but does not have as fast a growth rate. It has a gray to light brown face and a distinctive topknot of wool on its head.

## Texel

The Texel (Fig. 4.10) is smaller than the other ram breeds and is white-faced, with a slower growth rate. This breed has exceptional amounts of muscling. The Texel originated in the Netherlands and was imported in 1985.

## GENETIC TRAITS

It is not practical to try to improve a flock, especially a small one, in more than two or three traits at a time. There are three primary hereditary traits that concern the commercial producer: (1) general soundness, (2) multiple births, and (3) rate of gain. Depending on the nature of the operation, wool may also be a consideration. Soundness means conformation and the avoidance of certain defects. Multiple births and rate of gain are both critical factors in the profit level of an operation.

How much can the shepherd affect the flock through selective breeding, and how much is determined by environmental factors? It is not easy to answer these questions. Identifying environmental and genetic influences and trying to understand the relationship between them is a great ongoing scientific effort, but scientists have developed a way to let us talk about how much a particular trait may be influenced by heredity. They use the term *heritability* to indicate how much trait variation is controlled by heredity. Table 4.1 shows the average heritability for some sheep traits.

In a hypothetical flock of the same breed, for example, you will find

**4.8.** Hampshire. (*Courtesy, American Sheep Industry Association*)

**4.9.** Oxford. (*Courtesy, American Sheep Industry Association*)

**4.10.** Texel. (*Courtesy, American Sheep Industry Association*)

**Table 4.1.   Heritability of various traits**

| Trait | Percent |
|---|---|
| Reproductive | |
| Ewe Fertility | 5[a] |
| Prolificacy[b] | 10 |
| Scrotal circumference | 35 |
| Age at puberty | 25 |
| Lamb survival | 5 |
| Ewe productivity[c] | 20 |
| | |
| Growth | |
| Birth weight | 15 |
| 60-day weight | 20 |
| 90-day weight | 25 |
| 120-day weight | 30 |
| 240-day weight | 40 |
| Preweaning gain: birth–60 days | 20 |
| Postweaning gain: 60–120 days | 40 |
| | |
| Carcass | |
| Carcass weight | 35 |
| Weight of trimmed retail cuts | 45 |
| Percent trimmed retail cuts | 40 |
| Loin eye area | 50 |
| 12th rib fat thickness | 30 |
| Dressing percent | 10 |
| | |
| Fleece | |
| Grease fleece weight | 35 |
| Clean fleece weight | 25 |
| Yield (%) | 40 |
| Staple length | 55 |
| Fiber diameter | 40 |
| Crimp | 45 |
| Color | 45 |

*Source:* From *Sheep Production Handbook,* American Sheep Industry Association, Englewood, Colorado, 1992. (With permission)
[a]May increase to 10% in ewe lambs and in ewes bred in spring.
[b]Lambs born per ewe lambing.
[c]Pounds of lamb weaned per ewe exposed.

mature ewes of different weights. The average heritability tells us that these differences are, on the average, 40% the result of heredity. The rest of the differences could be due to environmental factors such as diet.

Some of the traits that make a sheep unsound are undershot and overshot jaws, inverted eyelids, horns on breeds that are supposed to be hornless, and a tendency to develop a rectal prolapse. Animals with these traits should not be used for breeding stock. Little is known about the exact degree of heritability in these traits.

Of the two main traits related to the profit margin of the flock, multiple birth and rate of growth, multiple birth is the most difficult to understand. Just because a ram and a ewe are both twins does not mean their offspring will be twins. With an average heritability of 15%, technically, twin birth is not highly influenced by genetics. Sheep have twins because of multiple ovulation, which is strongly influenced by the age of the ewe, the breeding time, and the nutritional state of the ewe. In addition to heritability, the amount of variation for a trait among animals is a factor in determining rate of genetic improvement—the greater the variation, the greater the progress. Since most flocks will have ewes that vary for the number of lambs born (from 1 to 3 or more), lambing rate shows a high variation. Selection of replacement ewes with a history of twin and triplet births will result in reasonable amounts of genetic improvement in the lambing rate.

Because of the complexities of genetics as a science, you can't use heritability to make hard-core predictions. However, it is useful to note some practical examples of how selecting for multiple births can influence flock production. Average heritability of 10% means that one could expect a slow but positive improvement in twinning by selecting ewes for breeding who are themselves twins. A yearly 2% increase in lambing can be accomplished by selecting twins for breeding. The more records kept on the ewe, the more accurately her genetic makeup can be appraised. Obviously, selecting ewes with several years of twin births is better than selecting ewes with only 1 year of twinning. Using a highly prolific breed like the Finnsheep in a cross-breeding program might accomplish in one generation what a lifetime of selection for this trait might accomplish in other breeds.

The late Paige Walters was a longtime Iowa Suffolk breeder who was well known for her meticulous breeding records. She offered the beginner the best reason for being concerned about the genetics of twinning. When asked what the heritability of multiple births meant to her, she replied quickly, "That twin lamb is 100% profit."

Rate of gain is more highly influenced by heredity than multiple births and is easier to chart because it can be measured on a continuous scale. By switching from a mediocre ram to a performance-tested, fast-growing ram, you might see improvement in rate of gain in the first year. Rate of gain is considered to be of medium heritability, but improvements can be quite dramatic. For example, if ram A gained 1 pound a day during the first 120 days, and ram B gained 0.85 pounds a day in the same period, the offspring from ram A can be expected to weigh 3–4 pounds more at 120 days than those of ram B.

## BREEDING PROGRAMS

### Market Lambs

As described earlier, the traditional commercial combination is to breed crossbred whiteface ewes that demonstrate superior maternal traits with a big, fast-growing blackface ram or terminal sire. The offspring of terminal sires are not meant to be kept for breeding. The bulk of income for the commercial sheep producer comes from sales of market lambs; therefore, a breeding program should be carried out with these sales in mind.

## Replacement Ewes

Ewes to replenish the flock or to increase flock size have become increasingly scarce and expensive. Many shepherds try to raise their own replacements. Earlier in this chapter, it was suggested that replacement ewes ideally would be sired by rams from one of the ewe breeds. However, when this is not practical, replacement ewes selected from a pool of lambs from whiteface ewes and terminal sires can make good-quality choices.

Many variations in breeding programs can be used, depending on the size of your operations. Flock owners can gain more genetic control by raising their own replacement ewes. In range flocks, a whiteface buck may be run with the ewes every third year and those ewe lambs kept as replacements. In an even more random way, owners may run both a blackface buck and a whiteface buck with the ewes at the same time. They would then save the best of the whiteface ewe lambs for replacements. Crossbreeding is important in your breeding program; improving a commercial flock without crossbreeding is slow and difficult.

## SELECTION OF REPLACEMENTS

It was mentioned previously that selecting against unsound features is the first step in choosing breeding stock. Inverted eyelids, prolapse, bad mouths, small testicles, ruptures, or faulty skeletal features should disqualify a ram or ewe immediately. Look at the growth rate of the remaining rams and the mothering traits (including multiple births) of the ewes.

The ram contributes to carcass quality and rate of gain. Of these traits, rate of gain is the most important. Because there is some positive correlation between the two, selecting for rate of gain will probably improve carcass quality. To find the rate of gain for a ram you must have a record of when it was born, if it was a twin or a single, how old its mother was, and whether she was a twin or a single. Weigh the lambs as close as possible to 90, 120 or 140 days of age.

Selecting ewes for multiple births is easier, but it produces less-dramatic results than selecting a ram for growing rate. Multiple birth is somewhat less heritable and its effects don't show up as quickly. If extensive flock histories were more readily available, the genetic potential of a ewe would be better known. For the most part, you will be restricted to choosing twin ewe lambs. Since older ewes are more apt to have twins than are ewe lambs or yearlings, a twin born of a ewe lamb or yearling might indicate a special tendency toward multiple births. Also consider the growth rate of these twin lambs. To select replacement ewes, list all of the sound twin ewe lambs from heaviest to lightest, according to their adjusted 120-day weight. Then make your selections, beginning with the heaviest. If too few sound twins are available, follow the same procedure for singles; choose the heaviest 120-day adjusted-weight singles. You can make adjustments so that all lambs in a group can be compared by using this formula:

$$120\text{-day weight} = \frac{\text{actual weight}}{\text{actual days of age}} \times 120$$

If you have a serious interest in improving your flock through genetics, the National Sheep Improvement Program (NSIP) may be of interest. It is a national computerized genetic evaluation program that calculates estimates of genetic value for every sheep in the flock for several maternal, growth, and wool traits. Flock owners provide the raw data and NSIP calculates the genetic values. NSIP is not a sheep management, ration-balancing, or sheep economics program—it is concerned with providing producers with information that will allow them to increase the rate of genetic improvement in their flocks.

If you enroll your flock in the NSIP, you will receive accurate estimates of an animal's genetic value for growth, maternal instincts, and wool quality, so you can identify superior animals. NSIP performs sophisticated genetic calculations that are not possible by hand or even on most personal computers. While it is relatively new, the program promises to be an important tool for genetic improvement. Your state sheep extension specialist can tell you how to enroll.

Your first impulse in selecting replacement breeding stock might be to choose the animal that looks the best. An understanding of the importance of breeding should help overcome this impulse. The only way to work at improving breeding is to keep records. This is something many shepherds, beginning and experienced, find odious. However, keeping and studying these records over the years can be a fascinating part of a sheep operation, and your time will be well spent.

# 5. Feeds and Feeding

*"The story of the ruminant is a beautiful story."*
—DR. CHARLES PARKER

*"A lot of people over feed and the sheep get too fat. But you have to be careful and remember how much you are asking the animal to produce."*—RANDY IRWIN

The sheep's ability to produce high-quality meat, wool, and milk from relatively low-cost feeds and forage has given them a special place in the history of humankind. Understanding how to support this efficiency through feeds and feeding is a key to both the profitability and health of the flock.

Sheep use proportionately more roughage than any other livestock. If cattle are marketed directly from pasture, they might not be as uniform in grade or as palatable. But sheep, when properly managed, can be finished directly from improved pasture or native range. Not every sheep producer has access to the pasture to do this, but it is possible.

Sheep are built for grazing. The sheep's muzzle is quite narrow, compared with the muzzle of a cow. The shape of the muzzle gives the sheep the ability to forage selectively. Sheep and cattle can make good grazing companions, since sheep prefer the short grasses and weeds (forbs) that cattle leave behind. Some studies have shown that sheep have an advantage over cattle in converting forage to meat. However, sheep are also efficient because they can use feed that might otherwise go to waste.

Sheep are ruminants, along with other cud-chewing animals such as cows, goats, and moose. Ruminants have a unique ability in the animal kingdom to process fibrous feeds, including stems, leaves, and seed hulls. Fiber is similar to other carbohydrates (starch and sugar) but more com-

**55**

plex chemically. Because of its complex chemical composition, fiber is practically indigestible for nonruminants.

Ruminants have a four-part stomach, but the unique food-processing abilities are located in the first two sections, known as the reticulo-rumen. Volumes have been written on the complex ecology of the ruminant stomach. But even the beginner needs to understand the basic workings in order to maintain the health and profitability of the flock.

Ruminants break down the complex carbohydrates in fibrous feeds through fermentation. Fermentation is the degradation of feeds by microbes in the rumen. The microbial population of the rumen is composed of several species of bacteria and protozoa, which are single-celled animals that are larger than bacteria. The microbes break down the feed for their own nutritional needs, and the byproducts of that process are what the sheep can use. When you feed your sheep, you are first feeding this complex population of microbes.

Each species of microbe is adapted to the fermentation of specific feed types, and each species of rumen microbe can degrade specific components of the feed. For example, only certain microbes can efficiently ferment fiber, while others can better use starches.

There are basically four sections of the ruminant digestive tract (Fig. 5.1): (1) the reticulum (honeycomb), (2) the rumen (paunch), (3)the omasum (many plies), and (4) the abomasum (true stomach).

Although these sections are identified separately, they actually constitute four compartments of one stomach. By contrast, most other mammals (such as humans and pigs) have only one stomach compartment.

When feed is first swallowed, it enters the reticulo-rumen, where it stays for a period of time, depending on whether it is roughage (high in fiber and bulk, such as hay) or a concentrate (relatively low in fiber and denser, such as grain).

From the reticulo-rumen, the feed is regurgitated and rechewed to produce smaller particles. Cud chewing probably evolved so ruminants could graze early or late while their predators were not hunting. Then, while they hid, they could be preparing their fiber feed for more thorough digestion. This first processing of feed after grazing, regurgitating, rechewing and reswallowing is when the fermentation process takes place.

Most protein consumed by sheep is degraded in the rumen. All protein is composed of nitrogen compounds known as amino acids. Both the sheep and the microbes in the sheep's rumen require specific amino acids. No single source of feed contains all known amino acids, but because the rumen microbes synthesize amino acids from other amino acids in the fermentation process, sheep rations do not need to be for-

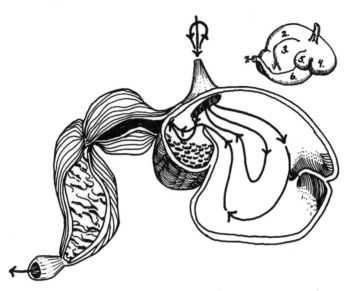

**5.1.** Ruminant stomach and cross section showing movement of food.
- 1—Esophagus
- 2—Dorsal sac of rumen
- 3—Ventral sac of rumen
- 4—Reticulum
- 5—Omasum
- 6—Abomasum

mulated to provide any specific amino acids as do swine and poultry rations. The microbes provide an additional supply of amino acids other than those in the feed eaten by the sheep.

Rumen microbes can degrade protein in feeds and utilize the ammonia, which is a product of this degradation, to synthesize amino acids. They combine the ammonia with carbon, hydrogen, and oxygen from the carbohydrate portion of the ration. Microbes can also utilize the ammonia released by the degradation of nonprotein nitrogen (NPN) sources such as urea and ammonium phosphate to make amino acids for their own maintenance and to supply microbial protein for the sheep.

Protein that escapes degradation by rumen organisms is degraded in the abomasum by the action of hydrochloric acid. The resultant free amino acids are absorbed by the abomasum in the same way that a stomach of nonruminant animals does.

The complex process of ruminant digestion is summarized in Figure 5.2.

## TAKING CARE OF THE RUMEN

Sheep *should* live primarily on roughage to maintain the proper microbial population in the rumen. The most common misunderstanding in feeding sheep is overfeeding of grain. Sheep don't need grain if high-quality roughage is available, and when grain is used, it must be used in proper balance to avoid health problems.

The species of microbes that utilize roughage produce byproducts that are essential to the sheep's growth. If these microbes are not available in sufficient supply, the sheep will be malnourished. It is important to change feeds gradually when changing a ration. This gives the specific microbes capable of fermenting the new feed time to increase, and for those microbes which cannot ferment the new feed to reduce.

For example, lambs are often removed from pasture (high-fiber diet) and their mother's milk and started on a ration containing grain (more starch, less fiber) for finishing. If this sort of change is made too abruptly, not only will the utilization of the new feed be inefficient because the mi-

**5.2.** The ruminant digestion factory. Microbes ferment roughage for their own maintenance → byproducts are fatty acids → used for energy in producing → milk, meat and wool.

crobe population is not in place, but toxic by-products and metabolic disorders such as enterotoxemia will occur. This is discussed in Chapter 6.

## NUTRIENTS

To make use of the sheep's ruminant efficiency, the proper balance of nutrients is needed. All feed is composed of nutrients, which are listed on the container to show the composition of the feed. Nutrient designations are used to determine the needs of the animal being fed. Matching feed nutrients with nutrient needs of the sheep is a major job of the shepherd.

The primary nutrients are protein, carbohydrate and fat (energy);minerals, both macrominerals and microminerals; vitamins; and water (essential, but not actually a nutrient). All feeds, grains and forage, have these nutrients in different quantities. Determining the proper proportions of these nutrients in a cost-effective manner is the challenge of developing feeding programs.

### Protein
The functions of protein include

1. Maintenance—Breathing, walking, grazing, finding water, chewing, repair of tissues, etc.
2. Growth—Growing lambs require a higher protein ration than adult sheep.
3. Reproduction—Gestation requires increasingly larger amounts of protein as fetal lambs develop; ovulation and sperm production require protein.
4. Lactation—Production of milk protein, casein.
5. Wool growth—The amino acids cystine and methionine are specific to wool and hoof growth and quality; sulfur is important in both of these amino acids.

### Carbohydrate and Fat
Energy is provided by the carbohydrate and fat in feeds; the carbohydrates are in the form of starch, fiber, and sugar. When carbohydrate is fermented, the products are volatile fatty acids (VFAs). These VFAs are a principle source of energy for ruminants.

Propionic acid is the VFA produced by the fermentation of grain and other feeds high in starch. It is used for daily energy and for lactation, with any excess stored as fat.

Acetic acid is produced by the fermentation of fiber. It is used principally for maintenance (heat increment) and can provide energy for growth.

The functions of energy include (a) daily maintenance, (b) growth, (c) ovulation (see "flushing," Chapter 7, p. 101), (d) lactation, (e) fat deposition (finishing lambs or dry ewes before breeding.)

Total digestible nutrients (TDN) is the measure of energy in a feed. TDN is a combination of digestible crude protein (DCP), digestible crude fiber (DCF), digestible ether extract (DEE), and digestible nitrogen-free extract (NFE), all which can be identified through standard National Resource Council (NRC) tables, or through individual analysis of feeds that you are using. TDN is expressed in units of weight or percent and is the most popular measure used for sheep ration formulation in the United States. The formula is

$$\%TDN = DCP + DCF + (DEE \times 2.25) + NFE \times 100$$

## Minerals

Minerals are present in all feeds in varying amounts but often must be supplemented for efficient performance. Minerals essential for animals are divided into two groups: macrominerals and microminerals (trace minerals). These names refer to the relative quantities of the minerals required, and their subsequent presence in animal tissue. The basic functions of these minerals are shown in Table 5.1. The NRC-recommended amounts of macrominerals and microminerals are shown in Tables 5.2 and 5.3.

**TABLE 5.1.   Basic functions of minerals**

| Mineral | Function |
| --- | --- |
| Calcium | Bone formation; muscle contraction; nerve function |
| Phosphorus | Bone formation; high-energy phosphate bonds; acid-base balance |
| Magnesium | Enzyme activator; bone formation |
| Sodium | Acid-base balance |
| Chlorine | Acid-base balance |
| Potassium | Osmotic pressure and acid-base balance; muscle activity |
| Sulfur | Tissue respiration; sulfur-containing  amino acids |
| Iron | Hemoglobin cytochromes; myoglobin |
| Copper | Hemoglobin synthesis; enzyme systems;  bone formation |
| Zinc | Enzyme systems |
| Manganese | Enzyme systems; amino acid metabolism; full-aid synthesis |
| Cobalt | Component of B12; promotes wool growth; prevents wool slippage |
| Iodine | Prevents goiter and woolless lambs |
| Selenium | Enzyme systems; enhanced immune function; works with Vitamin E |
| Molybdenum | Purine metabolism |

**Table 5.2. The National Research Council macro-mineral recommendations**

| Element | Recommended[a] |
|---|---|
| | (%) |
| Calcium | 0.20–0.82 |
| Phosphorus | 0.16–0.38 |
| Potassium | 0.50–0.80 |
| Magnesium | 0.12–0.18 |
| Sulfur | 0.14–0.26 |
| Sodium | 0.09–0.18 |
| Chlorine[b] | – |

[a]National Research Council recommended total ration nutrient density for sheep. The range is to allow for animals of different sizes and stages of production.

[b]Part of salt, which can be used to meet the sodium allowance.

**Table 5.3. National Research Council micromineral recommendations**

| Elements | Recommended[a] |
|---|---|
| Iodine | 0.10–0.80 ppm |
| Iron | 30–50 ppm |
| Copper | 7–11 ppm[a] |
| Molybdenum | 0.5 (max 10) ppm |
| Cobalt | 0.10–0.20 ppm |
| Manganese | 20–40 ppm |
| Zinc | 20–33 ppm |
| Selenium | 0.1 (max 2) ppm |

[a]Sheep are more sensitive to copper toxicosis than other mammals, so copper levels in minerals should be monitored carefully.

## Vitamins

Vitamins A, D, and E are classified as fat-soluble vitamins, and are seldom needed supplementally, except as "insurance" during late gestation and early lactation or when sheep are confined and fed harvested feeds.

Vitamin A is important to the health of the mucous membranes. Vision, reproduction, digestion, and lactation are all affected. It is also important for disease resistance, but an excess can be toxic. The adage that "if some is good, a lot must be better" does not hold true for Vitamin A. The toxic level is very high, so using moderation when adding Vitamin A will help avoid problems.

Vitamin D is synthesized through the action of sunlight on the skin. Confinement without sunlight, long wool, extended periods of cloudiness, or animals with black skin can make Vitamin D supplementation necessary. Vitamin D has a function in calcium metabolism, and a deficiency may result in lamb rickets. As with vitamin A, excess vitamin D supplementation may cause problems. The amount capable of causing these problems is extremely high and can result only from some sort of mistake or carelessness in mixing feeds.

Vitamin E is closely related to selenium in sheep nutrition, especially for fetal lambs and lambs less than 2 months old. Both white muscle disease and nutritional muscular dystrophy can result when lambs are deficient in vitamin E. Ewes can be supplemented before lambing with both selenium and vitamin E to aid in the control of the two disorders, especially in localities of known deficiencies.

The B vitamins are water soluble and can be synthesized by rumen microbes. However, they should be fed supplementally to lambs that were weaned before 2 months of age. Lambs without a functioning rumen apparently can't synthesize adequate levels of the B vitamins. Most grains are fairly high in B vitamins, but rations for young lambs seldom contain grain in an amount needed to provide B vitamin requirements. A lack of thiamin (vitamin B1) may be evidenced as a central nervous system disorder (staggering).

## Water

Sheep have the reputation of being hardy foragers able to go for long periods of time without water. While under stress, sheep can survive up to 7 days without water. However, a steady supply of clean ice-free water is an important part of the sheep diet. Water requirements of sheep vary with their reproductive and lactation status, the weather (temperature, wind velocity, humidity), wool (coat length and density), and feed (amount and composition). If allowed, sheep will consume water in an amount equal to two or three times their dry matter consumption on a weight basis. It is best to provide sheep of all classes with as much water as they will consume on a free-choice basis even though they can survive with access to water only once in 24 hours. As a rule of thumb, 1 gallon of water should be available for each 4 pounds of 90% dry matter consumed.

## DESIGNING FEEDING PROGRAMS

To raise sheep profitably, the shepherd must give sheep access to feed that meets their nutritional needs in a cost effective way. There are many information resources to help in this process, but ultimately, the shepherd must sit down with a pencil or a computer and analyze the options. Grain and supplements are usually considered the expensive part of a diet, making the sheep's need for roughage very cost-effective. However, as Stan

Potratz points out in Chapter 11, in some parts of the country, such as the Midwest, this is not always true. Using expensive crop ground for sheep pasture doesn't always make economic sense. In those situations, good management of pasture and making use of land that can't or shouldn't be tilled become especially important.

The NRC publishes feeding recommendations for all classes of livestock.(1) These recommendations are generally accepted in the U.S. livestock industry.

The availability and cost of various feeds can be obtained through your local feed sources and, of course, through your own inventory of homegrown feeds.

To design a ration, two categories of information are needed:

1. The weight (or age and weight), productive function of the animals to be fed, and recommended daily or total ration nutrient allowances taken from the NRC for that class of animal. The productive classes are open (neither pregnant nor lactating), bred or pregnant, lactating, and growing and fattening lambs.

2. A standard or specific nutrient analysis of available feeds, and the cost of these feeds.

As an example, we can use the NRC recommended daily nutrient allowances for feeding ewes weighing 150 pounds during the breeding season and for the last 4 weeks of gestation if a 180% to 225% lamb crop is expected (Table 5.4).

When designing a feeding program you must balance a ration using locally obtainable feeds (Table 5.5). During the breeding season, ewes

**Table 5.4.    National Resource Council recommended daily nutrient allowances (1)**

|  | Breeding | Pregnant |
|---|---|---|
| Dry matter (pounds/day) | 4.00 | 4.20 |
| Crude protein (pounds/day) | 0.25 | 0.47 |
| Total digestible nutrients (TDN) (pounds/day) | 2.30 | 2.80 |
| Calcium (grams/day) | 5.70 | 7.60 |
| Phosphorus (grams/day) | 3.20 | 4.50 |

*Source:* National Resource Council, *Nutrient Requirements of Sheep, 6th Edition,* Washington, D.C: National Academy Press, 1986.

*Comments:* [a] Although expected dry matter intake doesn't change much, recommended intake of all nutrients does increase, so nutrient density in rations should be raised in late gestation. [b] Tables of total ration nutrient density are available, but for hand-feeding of smaller groups of sheep, using daily nutrient allowances is more practical. To effectively use the tables of total ration nutrient density, mechanical mixing and feeding equipment is needed.

most likely get a good part of their nutritional needs from pasture, but for the sake of the formulation, alfalfa and earcorn are two commonly available feeds in many areas.

**Step 1:** First, average the dry matter (DM) content of the two feeds. Dry matter is the amount of a feed that is not water. It can be expressed as a percentage of a feed or as pounds that the animal needs. In this example, the alfalfa is 88% and the corn is 87%, which would mean 87.5% dry matter. From the NRC tables, the open ewe is expected to consume 4 pounds of dry matter. Since the DM percentage of the feed if 87.5%, you will need more than 4 pounds of feed to get 4 pounds of DM. To calculate this figure, divide the pounds needed by the DM percentage. This is the equivalent of +/− 4.57 pounds of the available feeds (4 divided by 87.5%).

**Step 2:** Next, you need to determine whether the feeds you have chosen can supply the amount of protein the ewe needs. To do this, divide the protein requirement (.25 pound) by the total pounds that will be fed (4.57 pounds), which shows the percentage of protein that must be in the feed. This calculation shows that a feed with 5.4% protein is needed. Since both feeds to be used contain more protein than this, protein is not a problem.

**Step 3:** The calculated weight of feed (4.57 pounds) must provide 2.3 pounds of TDN (see p. 60). To determine the TDN that must be available in the feed, divide 2.3 by 4.57. The feed should have a TDN of 50.3% to meet the nutritional needs. The TDN of the hay is slightly lower than this, and the TDN of the ground earcorn is higher, so it can be assumed that TDN allowance will be met with a small amount of earcorn with hay making up the rest of the ration. If the ration were 100% hay, it would provide 2.14 pounds of TDN (4.57 times 47% TDN). This leaves the ewe 0.16 pound of TDN short of recommendations. If 1 pound of hay is removed from the ration, and 1 pound of earcorn is added, the TDN intake will be increased 0.24 pound (0.71 pound of TDN added from earcorn, minus 0.47 pound of TDN deleted when hay is removed). To determine the exact amount of earcorn to substitute for hay, divide the 0.16 pound TDN deficiency by 0.24, the amount of TDN added by substituting 1

Table 5.5. **Feed nutrient analysis**

|  | Alfalfa/Grass hay | Earcorn |
|---|---|---|
| Dry matter (DM) (%) | 88.00 | 87.00 |
| Crude protein (%) | 11.00 | 7.80 |
| Total digestible nutrients (TDN) (%) | 47.00 | 71.00 |
| Calcium (%) | 0.65 | 0.06 |
| Phosphorus (%) | 0.21 | 0.24 |

pound of earcorn for 1 pound of hay. This shows that 0.67 pound of earcorn and 3.90 pounds of hay will provide the recommended allowances of protein and TDN to the ewe.

**Step 4:**   Calculate the amount of calcium provided by 3.90 pounds of hay and 0.67 pound of earcorn:

3.90 lb of hay × 0.65% (0.0065) = 0.0250 lb calcium +

0.67 lb of earcorn × 0.06% (0.0006) = 0.0004 lb calcium

Total calcium = 0.0254 lb × 454 (g/lb) = 11.53 g

**Step 5:**   Calculate the amount of phosphorus provided by 3.9 pounds of hay and 0.67 pound of earcorn.

3.90 × 0.21% (0.0021) = 0.008 lb phosphorus +

0.67 × 0.24% (0.0024) = 0.0016 lb phosphorus

Total phosphorus = 0.01 lb × 454 (g/lb) × 4.54 g

The ration provides an adequate amount of nutrients for the class of ewe (dry, breeding) it is intended to be used for. The same procedure can be followed in designing rations for other classes of sheep.

Tables 5.6 and 5.7 show typical rations for a ewe and a feeder lamb, respectively. These rations will supply recommended nutrient allowances, if trace mineralized salt is provided on a free-choice basis. When grain is recommended, it need not be processed for sheep.

Actual consumption may vary, but these proportions can be used.

## PASTURE/CROP AFTERMATH

Sheep of all classes can utilize pasture. Ideally pasture should form the basis of any sheep enterprise. If pasture is not adequate for all sheep, consideration should be given to eliminating or reducing the size of the sheep operation. While pasture should be of central importance to your operation, it is also the most difficult to assess. The many variables—type of pasture, moisture content, height of vegetation—make it difficult to establish nutritional recommendations.

Pasture may consist of native forage plants or planted grasses and legumes. Browse (brushy plants, woody plants, and shrubs) are utilized by

**Table 5.6.** **Example ewe ration (+/− 150 pound weight)**

| Kind of Feed | A | B | C | D |
|---|---|---|---|---|
| **Legume hay** | | | | |
| 1st 100 days bred | 4.50 | 1.50 | 0.00 | 0.00 |
| last 6 weeks bred | 5.00 | 1.50 | 0.00 | 0.00 |
| lactating | 5.00 | 1.50 | 0.00 | 0.00 |
| **Grass hay** | | | | |
| 1st 100 days bred | 0.00 | 0.00 | 4.50 | 0.00 |
| last 6 weeks bred | 0.00 | 0.00 | 4.50 | 0.00 |
| lactating | 0.00 | 0.00 | 5.00 | 0.00 |
| **Silage** | | | | |
| 1st 100 days bred | 0.00 | 5.00 | 0.00 | 9.50 |
| Last 6 weeks bred | 0.00 | 6.00 | 0.00 | 10.00 |
| lactating | 0.00 | 6.00 | 0.00 | 11.00 |
| **Protein[a]** | | | | |
| 1st 100 days bred | 0.00 | 0.00 | 0.25 | 0.25 |
| last 6 weeks bred | 0.00 | 0.00 | 0.33 | 0.33 |
| lactating | 0.00 | 0.25 | 0.50 | 0.66 |
| **Grain[b]** | | | | |
| 1st 100 days bred | 0.00 | 0.00 | 0.00 | 0.00 |
| last 6 weeks bred | 0.50 | 0.50 | 0.50 | 0.50 |
| lactating | 0.75 | 0.75 | 1.00 | 1.00 |
| **Mineral[c]** | | | | |
| all ewes bred/lactating | − | − | 0.03 | 0.03 |

Note: Rams during the breeding season, and for 2 or 3 weeks before, may be fed rations recommended for ewes, with larger amounts made available, depending on body weight and appetite of the rams. When not breeding, rams may be held on pasture or harvested roughage.

[a]Soybean meal, cottonseed meal, linseed meal.

[b]Corn, barley, wheat, oats, sorghum.

[c]Rations without legume hay should be supplemented with limestone or dicalcium phosphate mixed with salt and fed free choice. The half ounce intake shown above may or may not be consumed. Trace mineralized salt should also be fed free choice separately from the mineral/salt mixture.

sheep during the time of year when new shoots and leaves are appearing. Sheep will also consume broad-leafed plants classified as "weeds" (forbs).

The right stocking rate, or animals per unit of land, is important for maximum utilization with minimum supplemental feeding. When deciding how many animals a unit of land can carry, consider the fact that forage plants are seasonal, with periods of dormancy (perennials) or seasons when plants are dead without leaves or seeds (annuals).

Early-season (spring) forage plants can be characterized as being higher in all nutrients and lower in fiber than more-mature plants at the end of the growing season and after the season. Young pasture plants are relatively high in moisture, therefore, young, smaller sheep do not have the capacity to consume enough for maximum performance. In order to maintain efficiency, it may be necessary to provide supplemental feed

**Table 5.7.   Example feeder lamb rations (pounds per day)**

| Kind of Feed | A | B | C |
|---|---|---|---|
| Legume hay | | | |
| 60-lb lambs | 1.30 | 0.00 | 0.65 |
| 90-lb lambs | 1.80 | 0.00 | 0.90 |
| 110-lb lambs | 2.20 | 0.00 | 1.10 |
| Grain | | | |
| 60-lb lambs | 1.30 | 1.20 | 0.00 |
| 90-lb lambs | 1.80 | 1.90 | 0.00 |
| 110-lb lambs | 2.20 | 2.30 | 0.00 |
| Grass hay | | | |
| 60-lb lambs | 0.00 | 1.20 | 0.00 |
| 90-lb lambs | 0.00 | 1.90 | 0.00 |
| 110-lb lambs | 0.00 | 2.30 | 0.00 |
| Protein | | | |
| 60-lb lambs | 0.00 | 0.15 | 0.07 |
| 90-lb lambs | 0.00 | 0.15 | 0.11 |
| 110-lb lambs | 0.00 | 0.20 | 0.13 |
| Silage | | | |
| 60-lb lambs | 0.00 | 0.00 | 1.90 |
| 90-lb lambs | 0.00 | 0.00 | 2.70 |
| 110-lb lambs | 0.00 | 0.00 | 3.20 |

Note: [1] Ration C is best used for starting lambs on feed. When consumption levels off, it is probably more efficient to use feed proportions of one of the other rations. [2] Lambs fed ration B or C should have free choice limestone/salt available. [3] All lambs should have access to free choice trace mineralized salt.

during the late pasture season and after the pasture season (winter).

Most mature pasture is low in protein, so supplementation with one of the oil meals (soybean, cottonseed, or linseed) will not only provide needed protein, it will also improve the digestion of the mature pasture feed. Except under extremely poor pasture conditions, about a quarter pound of a supplement containing one-third oil meal will provide proper protein supplementation. If the condition of sheep and pasture warrants it, sheep should be fed roughage. Legume hay can be fed without supplement, but grass hay or silage (other than alfalfa haylage) should be accompanied by protein and calcium supplementation.

Remember, rumen microbes have a protein requirement and can ferment other feeds more efficiently if they have an adequate supply of protein.

No attempt will be made here to describe pasture evaluation for carrying capacity, because of the wide variation in soil fertility, climate (length of growing season), quality of forage, and type of sheep being pastured. However, observation of local sheep enterprises similar to yours and experience will aid in deciding stocking density.

Crop aftermath, such as cornstalks and hay stubble, makes a good source of maintenance feed for mature ewes that are not bred and are not lactating; however, it may not be nutritionally adequate without supplementation for growth of lambs. These roughages are similar to dormant perennial grasses or dead annual grasses, so they should be supplemented with protein. For dry roughage such as this, calcium is not usually deficient, but phosphorus usually is. All the oil meals supply phosphorus as well as protein.

Grazing sheep, if they are supplemented, should be counted and fed a measured amount of the supplement by hand to avoid enterotoxemia due to the consumption of too much nutrient-dense supplement. Also, the amount of supplement should be gradually increased over a period of several days, with plateaus at intervals for the purpose of adjusting the rumen microbial population to the supplement.

## FEED COSTS

It is important to make sure that your animals are getting the proper amount of all the nutrients to support their growth and maintenance. However, cost of feed will likely be your biggest production cost. There are many feeds and feed combinations that can meet the needs of your animals, so it is a good idea to analyze local feed availability and price, as well as your own resources, and chose an option that is most economical. A general formula for analyzing the costs of various elements of the ration is

$$\frac{\text{Cost per unit of nutrient}}{\text{Unit of weight} \times \%\text{concentration}} = \text{cost of feed per weight unit}$$

For example, a soybean meal costs $180 per ton and is 44% crude protein. What would you be paying per pound for crude protein? You would multiply 2,000 pounds by .44. Then divide the cost per ton, $180, by this amount. In this case the cost of the crude protein is $0.20 per ton. You could compare this with other protein sources to find the best value. The same simple formula will work for other elements of your ration.

Body condition scoring is a management tool that is being used to make sure that ewes are not overfed or underfed and to assess the readiness of lambs for market. It is done by feeling by hand for muscle and fat over the spine in the loin region. Condition scores range from 0 to 5, with 0 being very thin with no fat cover and 5 being overly fat. A condition score of 2.5 at the end of lactation and 3.5 at lambing is recommended. The ability to do condition scoring is a very useful management skill. The

*Sheep Production Handbook*(2) provides a more detailed description, but guidance from an experienced producer is important in learning this skill.

The NRC requirements and standard feed analysis are invaluable tools for you in planning your feed program. However, as in all aspects of shepherding, your own observations are also critical. Local soil conditions have an impact on the nutrient value of feeds. You may have bought sheep that were out of condition and need to catch up. Heat or cold can affect nutritional needs as well. The way your sheep look and act and their health status are factors you must take into account to make sure your animals are healthy and productive.

## REFERENCES

1. National Resource Council. *Nutrient Requirements of Sheep, 6th Edition.* Washington, D.C.: National Academy Press, 1986.

2. American Sheep Industry Association. *Sheep Production Handbook.* Englewood, CO: ASIA, 1988.

# 6. Health Care

*"We lost a lot of ewes. It was discouraging. We thought we were killing them."*—BILL BUTLER

*"As early as second grade, our kids would be making decisions about when to call the veterinarian."*—TOM AND SANDY CLAYMAN

> **Veterinary services, procedures, biologicals, and drugs mentioned in this chapter are not intended as recommendations without the consent of the producer's own practicing veterinarian.**

"A sick sheep is a dead sheep." At some point in your introduction to shepherding, someone will share this common *unwisdom* with you. Someone also will probably tell you not to bother to call a veterinarian for a sick sheep, either because the animal is so fragile it will die anyway and your money will be wasted, or because the local veterinarian has shown little interest in caring for sheep.

The health problems of sheep are as manageable as the health problems of other livestock, and beginning shepherds can manage most of the flock health needs on their own. The flocking instincts that make sheep relatively easy to handle and care for, however, do contribute to their reputation as bad health risks. These instincts are so strong that the sick sheep will struggle against illness to keep up with the flock. By the time the casual shepherd notices that a sheep has dropped out of the flock, it may be too late to save the animal's life. Attentive shepherding is the key to flock health.

Early in your efforts, you should establish a relationship with a local veterinarian. Sheep play a minor role in the practice of most veterinarians, so you may need to encourage them to offer their best services. All veterinarians are trained in disease diagnosis, to know the difference between sick and healthy cell tissues, to know about parasite cycles, and to know the effectiveness of various drugs. Their training is as useful with sheep as it is with other livestock. Even if you do not have a health crisis early in your shepherding, you can establish a working relationship with your veterinarian by asking for training in the proper use of a drenching gun, how to trim hooves, or how to dock and castrate your sheep. You will then have someone to turn to when a crisis does occur.

Even if an animal dies, it is often worthwhile to have the animal "posted," or necropsied, by your veterinarian. This can be an important investment in the health of the rest of the flock. It is important to take action before an individual problem becomes a flock problem.

In addition to sometimes having to deal with reluctant veterinarians, the conscientious shepherd is at a disadvantage when it comes to the availability of health care products. Sheep production is a small business in this country compared with other livestock, and drug companies therefore are not as committed to the industry. Getting FDA approval to market a product involves millions of dollars in field-testing and research, and when the patient population is relatively small, it just isn't worth the investment.

Many drugs and vaccines used in other species in this country are not approved for use in sheep in this country although they are in other countries. That is why it is sometimes necessary to establish a veterinarian-client-patient relationship—so these drugs can be used in your operation. Extra label drug use (ELDU) is explained later in this chapter.

This chapter covers the health problems common to flocks across the country. Most health problems can be prevented or controlled by daily observation, a sound program of disease prevention, and timely use of your veterinarian. Books and resources listed at the end of the chapter will help with your health care management.

## GENERAL HEALTH PROBLEMS

### Predators

While they are not a disease problem, predators are the largest cause of sheep deaths in this country. They are a major problem in the large flocks of the western states, and producers, environmentalists, and animal rights advocates are in debate about the best methods of control. Even in the farm-flock states, where sheep are apt to be kept close to a homestead, predators are still a problem. Coyotes or packs of stray dogs have done enough damage to discourage many farmers from keeping a small flock. A number of things can be done to help the problem. Some aggressive herd dogs, such as the Komondor and Great Pyrenees, can help. Electric fencing and corralling sheep at night may also help. The predator problem in your area can be assessed by consulting with extension personnel, conservation officers, and local sheep producers.

## Internal Parasites

Internal parasites are a major cause of death in sheep. Poor growth rates and susceptibility to other health problems are also directly related to internal parasites. The farm flock is especially prone to parasite problems because of close confinement.

The signs of serious parasite infestation are not always obvious. Waiting for obvious signs may often mean that treatment will be too late. Severely weakened sheep may die from an ordinary worming treatment and should be handled by a veterinarian. The signs of advanced parasite infestation include thinness, listlessness, diarrhea, and anemia, which is indicated by paleness of the inside of the eyelids. A swelling under the jaw, called bottle jaw, is another sign of serious problems.

The most effective treatment for internal parasites is a regular routine of worming. Pasture rotation, so that sheep are not always feeding on infested ground, is also important in managing the problem. Feeding sheep in a sanitary way, not from the ground or from feeders with droppings in them, is also important.

Your veterinarian can help you assess parasite problems in your flock through fecal analysis. Sheep have several kinds of worms, and this analysis will tell you the type and degree of infestation so that you can plan medication and treatment accordingly.

Most worms cannot be seen with the untrained eye; tapeworms are the exception. Liver flukes, flat oval worms, are a particular problem with western sheep. If you have purchased western sheep and have had a number of deaths among them, you should check their feces for liver fluke eggs or have a necropsy done. Tapeworms, however, tend to be an individual, not a flock, problem. Segments of the worms may be seen in mucus or in the feces.

There are four main ways to worm sheep: with a liquid drench given orally, with a bolus or pill, with injections, or with dry medication mixed in the feed.

In drenching, you must dose correctly. Dosing incorrectly can cause

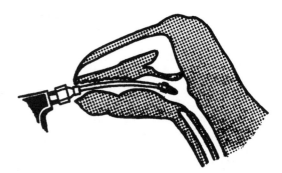

**6.1.** Correct dosing position (head in normal position).

liquid to go into the lungs, leading to pneumonia. To control sheep during dosing, stand behind the shoulder of the sheep. Hold the sheep under its jaw with one hand, clamping off the nose at the moment of dosing. Hold its head in a normal position, not twisted back or to the side. Insert the gun in the side of the mouth and over the tongue. Figure 6.1 shows how to position the gun during dosing. A veterinarian or experienced shepherd should help you do this the first time.

Pills or boluses are effective but sometimes difficult to insert properly. It is distressing, after giving medication to the flock, to find pieces of the boluses all over the ground. There is no way to know which sheep have been treated.

Effective and convenient injectable worming medications are on the market. The dose is important, for overdosing can lead to adverse reactions and even death.

The proper site and placement of injections are shown in Figures 6.2 and 6.3. For ewes and rams, use a 16-gauge 1/2-inch needle. The best place for subcutaneous injections is between the skin and the muscle in the lower rib area, avoiding the axillary space. Intramuscular injections are given in the muscle, preferably in the neck so that the prime cut will not be injured. Certain reproductive hormones for out-of-season breeding and rabies vaccine are the only products that should be given intramuscularly in sheep.

Some worm medications can be used in feed, as well. Sheep should have as uniform a dose as possible. Medications in feed are convenient, but some sheep may not get their share and, thus, not receive the proper dose. See your veterinarian for recommendations.

How often you worm your sheep will depend on many factors. If your sheep are on relatively damp pasture and not rotated often, they can accumulate parasites rapidly. If they are put on clean pasture after worming, control will be easier. Sheep being fed a relatively high level of grain tend

**6.2.** Subcutaneous injection sites and technique.

**6.3.** Intramuscular injection sites and technique.

to have fewer parasites than those on pasture. This is because the acid level in the stomachs of the grain-fed sheep is too high to support the parasites. Unless special circumstances dictate otherwise, worm in the fall when ewes come off pasture and again at lambing time or prior to going to pasture in the spring. In more moderate climates, frequency of worming (or deworming) must be increased and rotation becomes more important. Dewormers specifically effective against tape worms should be used if tapeworms are a problem.

## External Parasites

Sheep ticks, also known as sheep keds, cause sheep to chafe, lose wool, and gain weight slowly. The sheep tick is actually a wingless fly that spends its entire life cycle sucking blood from a sheep. When the adult sheep is shorn, the ticks settle onto the unshorn lambs for protection. A large population of ticks on lambs can produce restlessness and rubbing and chewing of the fleece. Expar is a product that can be poured on to treat ticks.

A common warm-weather problem for sheep is maggots. Various species of blowflies are attracted to damp areas on sheep caused by scouring (diarrhea) or cuts and sores. The flies lay eggs, which hatch into maggots that burrow into the flesh and cause severe damage and even death. In its advanced stages, maggot infestation is an unpleasant problem for the shepherd, as well as for the sheep. It can be treated easily in its early stages. Any sheep with damp or stained rear ends should be checked periodically for maggots. Animals that appear listless or that twitch around the tail should be checked too. Animals with maggots should be sheared around damp areas and treated with an effective insecticide. Likewise, any open sores or cuts should be examined and, if necessary, treated daily with an insecticide spray until the problem is controlled.

## Tetanus (Lockjaw)

Tetanus is a common, usually fatal, disease that is easily prevented. The anaerobic organism (*Clostridium tetani*) that releases the powerful tetanus toxin can live in the soil for years and may be present on any farm. Several years may go by without a case of tetanus in the flock, and then suddenly several sheep may die from it at the same time. Lambs that are banded for docking or castration are particularly susceptible. Tetanus usually enters the sheep through a wound, and once infected, sheep are unlikely to recover.

Three medications are used to control tetanus: tetanus antitoxin, tetanus toxoid, and penicillin. A combination of tetanus antitoxin and penicillin is effective in preventing tetanus at the time of docking and castration. Tetanus toxoid stimulates lambs to develop immunity. This requires two injections 2–3 weeks apart. Tetanus antitoxin is also used and gives immediate protection unlike tetanus toxoid. Lambs can be protected with 150 units of tetanus antitoxin if this is given early in life. However, the protection from antitoxin is of short duration, so lambs that you are keeping as replacements should be vaccinated twice. This may not be cost effective. Tetanus toxoid is also available in combination with products to prevent overeating disease. If tetanus is on your farm, immunity can be achieved with these combinations. The practice of vaccinating ewes with a tetanus toxoid prior to lambing is acceptable, but it is uncertain whether the lamb will be protected.

## Enterotoxemia (Overeating Disease)

Enterotoxemia, like tetanus, is caused by bacteria that produce a powerful poison. The bacteria, *Clostridium perfringens* type D, are normally present in the gastrointestinal tract of sheep. When the condition of the tract changes, the bacteria may multiply rapidly, producing the toxin that causes enterotoxemia. Enterotoxemia can lead to death in any sheep, but heavily nursing lambs and lambs starting on grain or lush pastures, or on feed for extended periods of time, are particularly susceptible.

Deaths from enterotoxemia are quite sudden. You may not see any sick animals and suddenly find one or more of them dead. In some cases, sheep may have convulsions. Some sheep with enterotoxemia develop diarrhea; these are more likely to recover. Enterotoxemia should not be confused with acidosis, an acid overload from ingesting too much grain.

The shepherd can control enterotoxemia through good management. Introduce sheep slowly to grain and control the grazing of harvested fields. If you graze sheep on a field following harvest, make sure they are vaccinated for type D enterotoxemia. Even though they are vaccinated, adjust them gradually to the feed, in case there is a lot of grain on the ground. You can control their foraging in two ways: limit the time on forage, starting with a half hour and gradually increasing the time, and limit grazing to small areas using portable electric fence.

As with tetanus, there is both a long-term toxoid and a short-term antitoxin available for enterotoxemia. Your veterinarian should be able to supply you with a combination enterotoxemia C and D/tetanus product. Ewes need two shots of toxoid the first year and a booster shot about 2 weeks before lambing to protect the newborn lamb. Lambs need one shot at about 5 weeks of age and a second 2–4 weeks later. If you buy lambs, vaccinate them immediately and then again 2 weeks later. Using antitoxin to treat sick animals is not very effective and is not cost effective.

## Pneumonia

Pneumonia is another common disease that can be partially controlled through proper management. Pneumonia is generally caused by *Pasteurella* bacteria or a combination of bacteria with respiratory viruses. Many environmental factors, such as stress caused by shipping, weather changes, castration, or weaning, increase the chance of pneumonia. Crowding, unclean bedding, moisture buildup, and drafts are other contributing factors.

Good management systems are the key to control. Ventilation is a major factor. There are breed differences in susceptibility to pneumonia, with the meat breeds being the most susceptible. Presently, there are no effective vaccines for prevention. Early diagnosis and sustained treatment with antibiotics and sulfas is essential in clinically sick animals.

Pneumonia may be hard to detect because it resembles other ailments. Signs of pneumonia include runny noses, listlessness, and lack of appetite. You may also hear a rattle in the chest of a newborn lamb with pneumonia.

## Footrot

Even though footrot is not fatal, it is a major cause of financial loss to sheep producers. Sheep with footrot cannot feed well and lose condition because of their lameness; rams may become too lame to breed, and lambs may take a long time to reach market weight. They become expensive to raise. Because footrot is highly contagious, it may affect a large part of the flock. A great deal of time is required to treat this disease, which puts more people out of the sheep business than any other.

Lameness is one symptom of footrot, although not all lameness is caused by this disease. The area between the toes becomes swollen, red, and moist. If not checked, the disease spreads and causes a separation of the hard hoof from the tissues underneath. The flesh gets soft, oozes, and has a strong putrid odor. In warm weather, maggots may infect the rotten flesh. If the feet are not trimmed, the hoof walls grow over the affected area and deform the feet.

Prevention is the most effective way to deal with footrot. Carefully examine new sheep and separate them from the rest of the flock for 30 days to make sure they do not harbor the disease. Treat the feet before they enter the flock. Regular hoof trimming is a critical part of control, because the bacteria that causes the disease, *Bacteroides (Fusobacterium) nodosus,* is anaerobic, and it thrives on the moisture and lack of oxygen under the folds of untrimmed hooves.

Treatment for footrot begins with a thorough trimming of the feet. Hoof trimming should be done regularly, whether footrot is present or not, but signs of footrot make trimming mandatory. To get the sheep in position for trimming, catch it by its rear flank; place your left hand under its jaw; and with your right hand at its dock, press the sheep firmly against your legs. Hold its lower jaw tightly and turn the sheep's head sharply over its right shoulder (Fig. 6.4). Press the right hand down on its flank and force the sheep to drop down.

Move your feet under the rump, raise its head, and position the sheep

**6.4.**   Taking sheep down for hoof trimming. (*Courtesy, Iowa State University Cooperative Extension Service*)

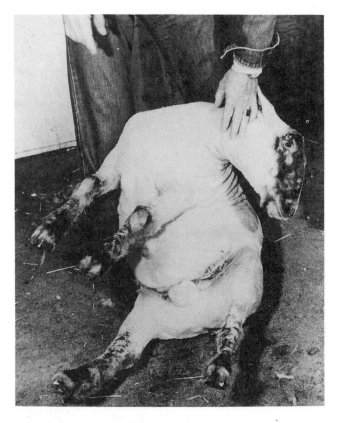

**6.5.** Position of sheep for trimming. (*Courtesy, Iowa State University Cooperative Extension Service*)

between your legs (Fig. 6.5). With your hoof-trimming shears, trim the hoof until it is level with the sole of the foot. If the toes are long and curled, trim those off carefully. Mark trimmed sheep with sheep-marking chalk so that you do not catch the same one twice.

During regular trimmings, those sheep with signs of footrot will need more care. You may find a sharp knife more effective for trimming sheep with footrot (Fig. 6.6). First, pare off all the ragged or separated hoof and sole to uncover infected areas; probe carefully for hidden pockets of infection. Sometimes you may need to cut healthy tissue to reach infected areas. Disinfect the knife and shears by washing with soap and water and dipping them in disinfectant. Vaccines are available that help prevent, though not totally, and control footrot.

Those sheep with active footrot should be separated from the rest of the flock for more extensive treatment, although even the healthy animals should be walked through a 10% formaldehyde footbath. When using formaldehyde baths, do not use at any less than 10-day intervals. Be sure feet are trimmed first because formaldehyde hardens the hooves. A topical preparation such as copper naphtholate for affected sheep may also be used after the bath. Affected sheep should be walked through a footbath every 10 days, preferably in a way that will force them to stand in the bath for as long as 5 minutes. Footbaths and trimming tables can be purchased from most of the equipment companies, or you can make one yourself. The bacteria that causes footrot can live in wet soil, so treated sheep should only be put on uncontaminated pasture.

## Contagious Ecthyma (Sore Mouth, Orf)

Sore mouth is caused by a virus that causes lesions and scabby sores around the mouth. Sore mouth creates four problem areas: clinical sore mouth is a danger to lactating ewes, sheep with sore mouth are barred

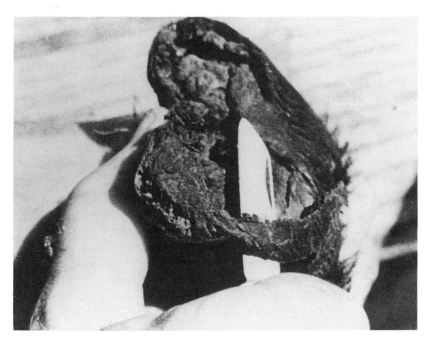

**6.6.** Trimming the hoof. (*Courtesy, Iowa State University Cooperative Extension Service*)

from exhibition, lambs on a lamb bar are particularly susceptible, and the shepherd can become infected.

Sore mouth can be transmitted to the udder by nursing lambs, leading to infections that damage the udder. Affected lambs will find it hard to nurse or eat and may lose weight. In severe cases, a lamb can die of starvation. A vaccine for sore mouth is available. Once sheep have had sore mouth, they will not get it again for some time, thus some think it makes sense to let the disease run its course if you are also able to protect the ewe from infection of the udder. The vaccine contains a live virus that can also cause lesions in humans. Treatment of affected lambs is unrewarding.

## Caseous Lymphadenitis

Caseous lymphadenitis causes abscesses of the lymph nodes that may go undetected until the sheep is slaughtered and federal inspectors condemn the carcass. Its symptoms are lumps, usually around the face and neck. Since the abscesses are walled off, systemic antibiotics are ineffective. The lumps need to be opened, drained, and treated with iodine or, if possible, surgically removed. The pus from the abscesses should be burned so that other sheep will not be affected. Affected sheep should be isolated until the wounds are healed. Internal abscesses sometimes cause wasting disease in breeding animals. A vaccine is available, but it may cause severe reactions.

## Plant Problems

Plants can harm sheep through both poisoning and mechanical injuries. The toxicity of plants varies from region to region and with the plants' stage of growth. Goldenrod, lupine, and poison vetch are three common plants that can poison sheep. As a rule, these plants will cause problems mostly when quality forage is not available. For example, goldenrod is most apt to be eaten during the winter, because it stays juicy when other forage is dry. The best way to prevent poisoning is to make sure that good feed is available and to supplement pasture with hay when it gets thin. Since the type of vegetation varies so much from region to region, the best way to become familiar with local poisonous plants is to check with your local extension service.

Weeds like foxtail, burdock, and cocklebur can also cause problems when sheep graze on them or when they are ingested with hay. They have seeds with barbs that can work into the membrane of the mouth and penetrate the skin. The seeds can also damage the quality of the fleece. These types of weeds can be controlled through proper grazing and irradication techniques suggested by local authorities.

# HEALTH PROBLEMS OF EWES
## Abortion

**Major Causes of Abortion.** There are four main types of infectious abortions in sheep: (1) enzootic abortion, (2) vibrionic abortion (*Campylocbacter*), (3) toxoplasmosis, and (4) salmonella abortion.

Toxoplasmosis and vibrionic abortion are the most common in the Midwest. Enzootic abortion may be more prevalent in the West but may become a problem in the Midwest in sheep that are imported from the West. Salmonella may occur sporadically anywhere at any time. Abortions due to listeriosis are usually seen in animals being fed silage.

> **WARNING: Some of the infectious agents that cause abortions in sheep are zoonotic. Pregnant women should stay out of the lambing barn.**

ENZOOTIC ABORTION.    Enzootic abortion in ewes (EAE) is characterized by abortions, stillbirths, and weak lambs. When EAE first appears in a flock, abortion rates may run from 25% to 60%. After the disease becomes endemic in an area, the incidence of abortion may drop to from 1% to 5%. The low incidence in the Midwest suggests that the disease is endemic there.

In the West, isolated range flocks become highly susceptible when the disease is introduced. Likewise, when western ewes are introduced into affected areas of the Midwest, they are very susceptible and high abortion rates may occur. Ewe lambs are most susceptible. The organism may cause pneumonia in young lambs. Initially, it was thought that the disease was spread only through contact with infected fetuses, placentas, or vaginal discharges, but it is now known that the disease is spread by infected sheep constantly shedding the infected agent in the feces or from the lungs.

VIBRIONIC ABORTION.    Vibrionic abortion is caused by the bacteria *Campylobacter fetus* or *Campylobacter jejuni*. *Campylobacter jejuni* is the predominant strain in the United States. Abortion rates are usually about 20% but may reach 80–90% in some outbreaks. Infected ewes generally recover following abortion and are immune to reinfection. Some ewes may remain persistently infected in the gall bladder and continue to shed bacteria in their feces. Some ewes die of complications such as an infected uterus, fetal retention, or peritonitis. Stillbirths and weak lambs are also common. Vibrionic abortion in sheep is not venereal. Ewes are infected by oral ingestion.

**TOXOPLASMOSIS.** Toxoplasmosis and vibrionic abortion are the most common causes of abortion in sheep in the Midwest. Toxoplasmosis is caused by *Toxoplasma gondii,* a protozoan that is spread by young cats that have eaten infected rodents. It comes from ewes ingesting feed or water that has been contaminated with cat feces. Toxoplasmosis generally does not cause clinical symptoms or detrimental effects in open healthy ewes. In stressed ewes and immunosuppressed ewes, neurological symptoms and death may occur.

The effect of toxoplasmosis in pregnant ewes varies with the age of the fetuses when the ewe is infected. Infection in the first 2 months of gestation results in embryonic death and reabsorption; infection in midgestation results in abortion and infection. Infection in the last trimester of gestation results in abortions, stillbirths, mummies, or weak lambs.

Flock abortion losses attributable to *T. gondii* can involve from 5% to 50% of the ewe flock, with typical losses averaging 15–20% of the lamb crop.

**SALMONELLA ABORTION.** Salmonella abortion is a distant fourth in frequency as a cause of abortion but probably occurs more often than recognized. The two major factors that determine whether a pregnant ewe will abort from *Salmonella* infection are stress on the ewe and the number of bacteria the ewe ingests.

Abortions may occur early in gestation but are most common in the last month of gestation. Abortion rates can approach 70%. Most of the ewes show diarrhea and some will die from metritis, peritonitis, and septicemia. Healthy lambs may also contract the disease and die.

**Controlling Abortion.** Both prevention efforts and aggressive treatment in the face of outbreaks are needed to control abortion.

With the similarity of symptoms, the time delay in establishing an exact cause and the possibility of mixed infections, it is critical to begin aggressive therapeutic regimens in an abortion outbreak.

**Prevention**

1. Develop an effective vaccination program. Vaccinate all ewes prior to breeding with a federally licensed, oil-based vibrio/enzootic abortion/*E.*

*coli* product. This three-way combination is currently the only available product for ovine abortion but should not be used for 30 days after breeding since the *E. coli* component may cause embryo loss. Ewes that haven't been previously vaccinated need a booster vaccination at midgestation. If you experience vibronic abortion loss, it may be important to have your veterinarian prepare an autogenous bacterin for your flock, one made from the specific strain of bacteria present.

2. Feed 200 milligrams of chlortetracycline per head per day during the last 6 weeks of gestation.

3. DO NOT feed on the ground or allow sheep to drink from stagnant fecal-contaminated pools.

4. Prevent contamination of feed and water with feces of rodents, birds, and cats. Neuter cats and maintain a stable adult cat population.

5. Maintain first-lambing ewes as a separate unit.

6. Maintain purchased replacement ewes as a separate unit.

7. Avoid stressing and crowding the sheep, and avoid unsanitary facilities.

8. Dispose of placenta and dead or aborted lambs immediately. Do not mix ewes that have aborted with other pregnant ewes. Do not bed pregnant ewes with bedding from the lambing area.

**Treatment of Outbreaks.**   Contact your veterinarian for assistance with an outbreak of abortion. The following are steps that have been used successfully.

1. Submit aborted fetuses and placentas to a diagnostic laboratory. Specimens should be sent to a laboratory that has experience in identifying the infectious agents. It is crucial that several specimens be submitted periodically throughout the abortion storm. This is necessary because more than one type of abortion may be present.

2. Develop an aggressive program of antibiotics in the feed with the help of your veterinarian.

3. Isolate aborting ewes from the rest of the flock.

4. Discontinue feeding on the ground and check for contamination of feed supplies.

5. If *Salmonella* is involved, base treatment on antibiotic sensitivity.

Vaccination with bacterins and the feeding of chlortetracycline are not the only approach or only options. Other approaches to prevention and control have been used with success, even though FDA-approved drugs are limited. If abortion is a problem in your flock, you need to work closely with a veterinarian and explore other treatment approaches. State extension service veterinarians can inform you about the nature of outbreaks in your area.

## Pregnancy Toxemia

Pregnancy toxemia, or lambing paralysis, occurs in the last month of gestation when the baby lamb gains approximately 80% of its birth weight. This growth causes a serious drain on the ewe. If the ewe is not getting sufficient nutrients, her lamb will develop normally, but she will draw on her fat resources to maintain herself. When the fat is broken down, ketone bodies toxic to the ewe are formed. She will get listless, lag behind the flock, and eat less—she may even die. The disease is most apt to occur in ewes carrying twins or triplets. For the average ewe carrying more than one lamb, 3 pounds of hay and 1 pound of grain per day may not be enough. Once the lamb is born, the ewe will probably recover; in some cases, inducing birth is a proper treatment.

At one time shepherds attributed pregnancy toxemia to a lack of exercise during pregnancy. It is now considered to be brought on by poor nutrition management, allowing ewes to become too fat in early pregnancy. Ewes must be carefully watched in late pregnancy. The sick ewe requires quick energy. Adding molasses to her feed is one way to do this. The sick ewe can be drenched with 2-3 ounces of propylene glycol administered orally three times daily or given dextrose intravenously, subcutaneously, or intramuscularly. For more-severely affected ewes, dextrose given intravenously is recommended. Vitamin B12 is sometimes used to increase the appetite.

## Vaginal and Uterine Prolapses

The tendency to have vaginal prolapses is hereditary. Prolapses usually occur 1 or 2 weeks before lambing. If the vagina moves in and out, you may simply treat it with a topical antibiotic to prevent infection. If it comes out and stays out and the ewe is straining, you may need to stitch it in. You can learn to stitch it yourself, but you may want to call a veterinarian the first time. A vaginal retainer can also be used for this problem. The prolapse should be cleaned well with soap and water and then put back in. If the prolapse is kept in with a stitch, the stitch should be removed within several hours of lambing. If a retainer is used, the ewe forces the retainer out during labor, or the lambs are delivered over it. The problem usually disappears after lambing, and the vagina stays in. Figure 6.7 shows the steps in treating vaginal prolapse with a retainer. Occasionally, the ewes need to be restitched.

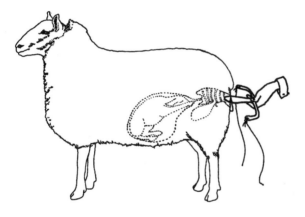

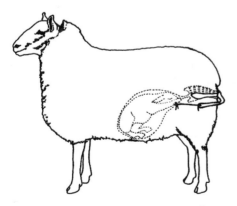

**6.7.** Vaginal prolapse and treatment with a retainer.

Uterine prolapse occurs after a lamb is born, most often if the lamb is large. The uterus must be put back in. Wash the uterus carefully, place antibiotic pills in the uterus, use oxytocin to shrink the uterus, and use plenty of lubricating material to facilitate the reinsertion. Stitches will hold the uterus in place until it returns to its normal size (Fig. 6.8). Use a mattress stitch with 3/8-inch umbilical tape behind the muscle in the vaginal wall. Antibiotics are often used to prevent infection. Often, after stitching, the ewe will continue to strain. Giving the ewe two aspirins two or three times daily helps alleviate this straining. A cortisone injection may also be helpful.

## Mastitis

There are two times when mastitis is a problem: after lambing and after weaning. The type of mastitis that occurs after lambing can be caused by *Staphylococcus* spp., *Streptococcus* spp., *Pasteurella* spp., and *E. coli* bacteria. The udder becomes swollen, inflamed, and hardened so that the lamb is unable to nurse. Good sanitation helps prevent this type of mas-

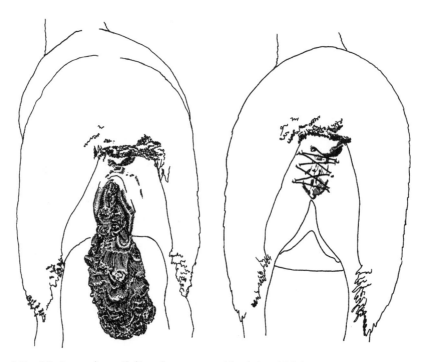

**6.8.** Uterine prolapse (*left*) and treatment with stitches (*right*).

titis. Hot pads or even hot towels on the udder may help. The shepherd should make sure the baby lamb is getting milk.

The second type of mastitis occurs when heavy-milking ewes wean their lambs. This is often caused by pasteurellae. The udders become swollen with milk. A systemic infection results, and the udder often becomes cold and blue. The ewes become sick, go off their feed, and may even die. This type of mastitis can usually be prevented through good management. The ewes should be kept off grain for a week before weaning, and alfalfa hay should be replaced by grass hay. If ewes are in poor condition, decreasing nutrition is not necessary. Water deprivation does not help.

If you see ewes with badly swollen udders or with a stiff, limping walk, they should be milked and treated for infection. The swelling of the udder may prevent antibiotics from working, in which case the udder itself must be infused with medication. Cattle mastitis tubes can be used for medication, using a tube on each side of the udder. Dry cow mastitis tubes can be used to infuse the udders of heavy-milking ewes 1 week following weaning, to prevent problems.

Mastitis of either type can make the ewe an unfit milker. If the udder is simply uneven, it may still be functional, but if it is hard too, the ewe probably will not be able to raise a lamb.

## Ovine Progressive Pneumonia

Ovine progressive pneumonia (OPP) is a wide-spread viral disease in adult sheep. Sheep can test positive for the virus with minimal or no signs of clinical disease. Eradication involves laboratory expenses and may also remove some of your productive ewes. The laboratory tests are not always accurate. Symptoms can include weight loss, difficulty in breathing, and development of lameness, paralysis, and mastitis. It can eventually lead to death. Lactation failure, referred to as hardbag, is related to the OPP virus and is being researched at several state universities. Many flocks that are testing positive, however, have no milking problems.

In a recent study (1), 16,827 sheep from 29 states were tested; 26% of the sheep tested positive and 48% of the flocks had positive sheep. With few exceptions, according to Randall Cutlip, DVM, sheep seroconvert within 2 months after experimental inoculation with the OPP virus.

Ovine progressive pneumonia virus is spread by dirty needles, by the feces, and ingestion of contaminated feed or infected cells in the colostrum or milk by the newborn.

Some breeds of sheep may be more susceptible to the virus. A higher level of clinical disease may be in Finnish-Landrace sheep and Texel sheep. Hampshire sheep seem to show a higher level of natural resistance.

Your approach to OPP must be part of your own personal long-term plan for flock health. Producers must determine the level of economic losses in their herd, for the costs of testing and eradication can be greater than the disease losses. Two popular tests, the agar-gel-immunodiffusion (AGID) and the enzyme-linked immunosorbent assay (ELISA), currently cost $5.00 per sample and should be done annually to obtain a high level of accuracy. Culling decisions should be based on ewe productivity in commercial herds. If you raise purebred sheep and want to sell OPP-free sheep for breeding, you may be able to justify additional costs. For the average commercial flock, it may not be economical to maintain an OPP-free flock.

## MANAGING COMMON HEALTH PROBLEMS OF LAMBS

How the lamb gets started is a big factor in susceptibility to disease. Most lamb deaths occur in the first few days of life and are related to the following diseases.

1. Starvation: Starvation remains the leading cause of death in young lambs. Making sure the teats are open, the milk supply is adequate, and the lamb has nursed are absolutely essential. Tubing of lambs that for various reasons haven't nursed is absolutely essential. An average lamb needs 50 ounces of milk in the first 24 hours of life. Don't underfeed.

2. Colostrum deprivation: Colostrum is the enriched first milk that the ewe produces. As well as providing antibodies, colostrum is a source of energy, and has a laxative effect as well as other unidentified factors. Failure to receive adequate colostrum results in lambs that are more susceptible to disease. Lambs deprived of colostrum seldom survive. Excess ewe colostrum should be frozen and thawed when needed. Do not use a microwave for thawing. Bovine colostrum may be substituted for water when making lamb milk replacer. A dried sheep colostrum milk supplement is now available.

3. *E. coli* scours: *E. coli* causes an acute diarrhea that occurs within hours of birth, generally resulting in death. It can be prevented by vaccinating the ewes with an *E. coli* product 4 weeks prior to lambing. If you have given the antiabortion vaccine containing *E. coli*, this should help. Antibody products can be given at birth for prevention, although this method is more expensive. In acute outbreaks, an antibiotic can be given orally to the lamb immediately after the first nursing. Once the lamb has

diarrhea, electrolytes and antibiotics are much less effective. Adequate intake of colostrum is important in prevention. Do not bed pregnant ewes with bedding from the lambing area.

4. Pneumonia: Pneumonia is generally not a problem at lambing time, but it occurs as lambs get older. Occasionally young lambs exhibit pneumonia-like symptoms when dehydrated from scours, loss of blood, or when there is a defective heart and lung function at birth. Vitamin E and selenium deficiencies will cause poor heart and lung function, making lambs more susceptible to pneumonia. Organisms that cause abortion in the ewes may produce damage to the liver, heart, and lungs, so lambs are born with pneumonia-like symptoms. Overcrowding and poor ventilation are the main contributing factors.

5. Vitamin E and selenium deficiency: Both vitamin E and selenium contribute to normal muscle function, and they interact with each other. To some extent they can replace each other, but not totally. Deficiencies result in poor heart and circulatory function and may also affect the muscles for breathing. Lambs may be born weak. As lambs get older and begin to grow rapidly with a deficiency, sudden death or white muscle disease may develop. Symptoms often develop after forced exercise. The most cost-effective and safest way to provide selenium is through feed or salt. Vitamin E can be in the feed, injected, or administered orally. Ensure a source of both prior to lambing. Baby lambs are born with no detectable levels of vitamin E, and are dependent on the ewes' milk for it. Particularly in early lambing, it may be advisable to supplement vitamin E to the lambs either orally or by injection.

6. Navel infection: Sanitation is important. Producers dab navels with iodine, but ½ to 1 cubic centimeter of penicillin at birth may be more effective.

7. Inverted eyelids: Affected lambs develop watery eyes soon after birth. The bottom lid is turned in and hair irritates the eye. This problem can be corrected with wound clips that are fast, easy, and economical to use.

8. Enterotoxemia: This problem appears as bloody diarrhea soon after birth or as sudden death in a solid and healthy lamb at about 3–5 weeks of age. Prevention is accomplished by vaccinating the ewe prior to lambing. Antibiotics and antiserum can be used to halt outbreaks.

## Pink Eye

Pink eye is common in young lambs and feeder lambs. Their eyes become cloudy and inflamed and the white part is tinted with red. Excessive tearing occurs. Pink eye is contagious and can spread easily through a flock. Antibiotics given by injection and ophthalmic preparations may help, but the disease runs its course in 2–3 weeks, whether it is treated or not.

## Urinary Calculi

Urinary calculi are most common in feedlot lambs, but other sheep may be affected as well. Stones form in the urinary tract. Affected sheep stand with their backs hunched, straining as they try to pass urine. The high concentration of phosphorus in a heavy-grain diet is thought to be a cause of this disease. The ration needs to be correctly balanced for calcium and phosphorous to prevent the formation of urinary calculi in the bladder. The total ration should be two parts calcium to one part phosphorous. Adding 30 pounds of limestone, 20 pounds of salt and 5–10 pounds of ammonium chloride per ton of feed will help prevent losses from urinary calculi. Affected animals can also be treated with an ammonium chloride drench. Surgery can relieve affected animals if they cannot otherwise be helped.

## Rectal Prolapse

Rectal prolapse is a frequent problem in many feedlot lambs. Many factors are thought to be associated with this condition including (1) coughing from respiratory infection or dust, (2) hereditary weakness, (3) docking tails too short, (4) growth implants, (5) fat deposits in the pelvic canal that come with overfeeding of grain and protein concentrates, and (6) straining from other disease conditions such as coccidiosis or urinary calculi.

Producers should consider slaughtering lambs that are near market weight. If treatment is selected, good results have been achieved by early detection and injecting a counter irritant, such as 7% tincture of iodine, in the tissue around the rectum. The counter irritant stimulates scar tissue, which eventually will help attach the rectum within the pelvic cavity. A purse-string stitch may also be used. Your veterinarian can demonstrate these techniques. If coughing is a problem, treating the lamb for upper respiratory disease is important.

## Epididymitis

Epididymitis is a bacterial infection that affects both ram lambs and adult rams. The epididymis is the tubular portion of the testicle that collects and stores sperm. The epididymis becomes swollen and the testicles may shrink. The disease is transmitted by rams. Rams are of low fertility after the infection. When you buy rams, the epididymis should be palpated for signs of the disease. Your veterinarian can teach you how to do this. A blood test can also establish the presence of bacteria.

## Coccidiosis

Coccidiosis is a very important disease in lambs. It may appear as a mucous-filled gray and light green diarrhea in nursing lambs 3 weeks and older. It is also a common disease in feedlot lambs. Prevention can be accomplished by providing additives in the feed.

Clinical treatment is best accomplished with sulfa drugs—sulfamethazine, sulfadimethoxine, and sulfaquinoxaline. Treating flocks of feedlot lambs by medicating their drinking water, and individual treatment of nursing lambs by drench are effective. Treatment is of questionable benefit in lambs showing clinical signs, so all lambs in the group should be treated even if not yet sick.

## THE SHEPHERD'S HEALTH

There are a number of sheep diseases that can affect the shepherd. These diseases include enzootic abortion in ewes, toxoplasmosis, salmonellosis, cryptosporidiosis, and sore mouth.

Enzootic abortion in ewes (EAE) is caused by *Chlamydia psittaci*. Whether sheep abort or deliver viable lambs, ewes with a placental chlamydial infection shed extremely large numbers of chlamydiae in the placenta and fluid discharges, which are in the air. Shepherds at high risk are those who lamb indoors and have enzootic abortions occurring in their flock. Pregnant women should take special precautions since *C. psittaci* seems to have an affinity for the human placenta and can provoke abortion or stillbirth. Clinically, the people will experience respiratory illness including pneumonia, muscle and joint pain, general malaise, and fever. Chlyamidial infections respond well to early treatment with tetracycline or other appropriate antibiotics.

Toxoplasmosis is caused by *Toxoplasma gondii*. This agent can infect virtually all warm-blooded animals. About 30% of the world's human population have antibodies to this agent, indicating a very high exposure level. *T. gondii* causes late-term abortion in sheep. The main people at risk are pregnant female shepherds. If they receive their first exposure to this disease early in pregnancy, the result is usually spontaneous abortion, stillbirth, or severe disease to the unborn child. Maternal infection in later pregnancy most commonly results in subclinically infected infants who may go on to develop more-severe symptoms such as mental retardation. The general symptoms of toxoplasmosis include malaise, fever, and lymphadenopathy. Again, it is recommended that the pregnant shepherd avoid working with sheep during the lambing period. In addition to this, when abortions occur, precautions should be taken, such as the wearing gloves, washing the hands, and using disinfectants. The treatments for toxoplasmosis infections in people include pyrimethamines and sulfon-amides.

Salmonellosis is most frequently caused by *Salmonella typhimurium.* In outbreaks the sheep are usually very depressed and have severe diarrhea and fever. Pregnant ewes will often abort. Mortality can often be quite high. Human infection results from contact with infected sheep and then with one's own oral cavity. The organism then grows in the intestinal tract and releases endotoxin, which causes the symptoms. The symptoms oc-cur 8–72 hours after ingestion and include diarrhea, nausea, abdominal pain, prostration, chills, fever, and vomiting.

Cryptosporidia is a protozoan that can also infect people. The most prominent sign in sheep as with any other species is diarrhea in infected lambs. The route of infection for people is through the ingestion of con-taminated feces. The predominant symptoms are diarrhea, abdominal pain, vomiting, anorexia, and fever. The incubation time is 3–8 days, and illness often begins with mild fever, nausea, and anorexia. This is followed by diarrhea. The severity of the diarrhea usually diminishes after the first week; however, it may last as long as 3–4 weeks. Human treatment for this disease is simply supportive in an attempt to maintain hydration. This disease is best prevented by practicing good hygiene.

Sore mouth is caused by the *Parapoxvirus.* The lesions commonly found in sheep include blisters that soon become scablike around the mouth and nose of lambs. These lesions may also be found on the udders and feet of ewes. In people the lesions are most commonly found on the skin of the fingers or thumbs. These lesions are usually raised, circular to oval, and about 0.5–1.5 centimeters in diameter. There are often enlarged regional lymph nodes and fever accompanying sore mouth. The lesions are painful. Repeat infections cause a raised, hard semicircular nodule somewhat smaller and less inflamed. These lesions are usually itchy rather than painful. Lesions generally resolve in 4–8 weeks. Treatment for orf in humans includes using tincture of iodine or ethyl ether.

Q fever is carried by sheep but causes no clinical disease. It can cause severe flu-like symptoms in humans that inhale the organisms.

## DRUG PRODUCTS

There are a wide variety of antibiotics on the market, only a few of which are approved for sheep. These need to be selected on the basis of the specific condition, with the help of your veterinarian. Pneumonias, abortion outbreaks, mastitis, and navel infections are some examples of conditions where antibiotic use is indicated. Table 6.1 provides a list of

**Table 6.1.    Approved drugs for use with sheep**

| Product | Route[a] | Dosage | Frequency[b] | Withdrawal (days) |
|---|---|---|---|---|
| Procaine Pen G | SQ | 3,000 IU/lb | SID | 9 |
| Erythromycin 100 | SQ | 1 mg/lb | SID | 2 |
| Erythromycin 200 | SQ | 1 mg/lb | SID | 10 |
| Terramycin | Eye ointment | | | |
| | | Internal Parasites | | |
| Ivomec drench | Oral | 3 ml/26 lb | | 11 |
| Tramisol drench | Oral | 1 ml/10 lb | | 3 |
| Tramisol bolus | Oral | 1 bolus/50 lb | | 3 |
| | | External Parasite Pour-ons (all 4 oz down midline) | | |
| Expar 11% EC | | 1 pt to 25 gal water | | |
| Expar pour-on | | ½ oz/100 lb | | |
| Ectrin spray | | 1 qt to 100 gal water | | |
| Ecrtrin pour-on | | 1 qt to 12½ gal water | | |

[a]IM = intramuscular; SQ = subcutaneously.
[b]SID = once a day, every 24 hr; BID = twice a day, every 12 hr; TID = three times a day, every 8 hr; 48 hr = every other day, 48 hr.

drug products with approval status, dosage, frequency of administration, and withdrawal times.

## SHEEP: EXTRA LABEL DRUG USE

One of the major frustrations facing sheep producers and their veterinarians is the lack of approved products for use in flock health programs. Many products are developed and available for use, but the Food and Drug Administration (FDA) has not approved their use in sheep. A producer should work closely with a veterinarian to establish a basis for extra label drug use (ELDU) when approved products are not available to treat a specific condition. Using any product that is not approved and labeled for use in sheep constitutes extra label drug usage. Much of the problem is due to the relatively small numbers of sheep in the United States. Drug companies cannot economically justify going through the testing required to approve a drug for use in sheep because they are considered a minor species. The National Research Support Program, through the FDA, currently has funds available for drug research in minor species such as sheep. The results are public property. This can make it more feasible for companies to go through the approval process. The Center for Veterinary Medicine (CVM) in the FDA has a drug-use guide for sheep and goats that is available.(2) They also have a home page on the World Wide Web with a searchable database on approved drugs.(3)

Several drugs, dewormers, and feed additives are approved for use in sheep and are effective products. Producers and veterinarians prefer using approved products with a known safety and proven efficacy. However, sometimes extra labeling a drug is justified and necessary to protect and promote the well-being of an animal. An unapproved product does not necessarily indicate that it is unsafe or ineffective, but rather that it has not been approved by the FDA for use in sheep.

Examples of ELDU by a veterinarian are (1) using a higher dosage than what is allowed on the label, (2) using the drug in a different species, or (3) using the product for a different indication than what it is labeled for.

Criteria for ELDU in food-producing animals have been established, and all the criteria must be met to be in compliance with the FDA. These criteria are (1) the practitioner must be a licensed veterinarian; (2) there is an established veterinarian-client-patient relationship; (3) there is an established medical diagnosis and need for treatment; (4) labeled treatments have been ruled out because (a) none are available or (b) they have been studied and/or used and are judged not to be the most effective under existing conditions; (5) treated animals are identified and records kept; and (6) the veterinarian is responsible for including withdrawal times and if necessary any follow-up treatment.

Veterinarians can adapt treatments available from the cattle industry. Cattle are a larger ruminant species and much research and product development has been done for cattle. Many of the antibiotics and dewormers cleared for use in cattle are effective in sheep but using them constitutes ELDU. When using a product that is not approved or labeled for use in sheep, extra precautions for withdrawal times and the possibility of adverse reactions must be taken. The ability to make use of ELDU is one of many reasons to have an established relationship with a veterinarian.

## YOUR RESPONSIBILITY

This chapter has discussed the most common health problems in sheep. The topics covered here are not all-inclusive. Your sheep may have a less common health problem, or there may be a local problem in your area that has not been covered. The reading list at the end of this chapter can provide you with more extensive references on health problems. Do not neglect flock health problems because of the sheep's reputation as a difficult and unresponsive patient. Watching your animals carefully and taking responsibility for timely diagnosis and treatment of problems, with the help of a veterinarian, will do much to improve this reputation.

## REFERENCES

1. Cutlip, R.C.; Lehmkuhl, H.C.; Sacks, J.M.; and Weaver, A.L. Seroprevalence of ovine progressive pneumonia virus infection in the United States. National Animal Disease Center.

2. Food and Drug Administration. CVM Memo: Drug Use Guide—Sheep and Goats. HHS Publ. No. (FDA)95-6019, June 1995.

3. CVM Home page: http://www_cvm.fda.gov.

## FURTHER READING

American Sheep Industry Association. *Sheep Production Handbook.* Englewood, CO: ASIA, 1988.
    Covers the basic areas of sheep production, nutrition, health, and management.
Henderson, David C. *The Veterinary Book for Sheep Farmers.* Farming Press Book, 1990.
    A 696-page, profusely illustrated volume aimed at helping the sheep farmer increase the productivity and profitability of the flock.
Linklater, K.A., and Smith, M.C. *Diseases and Disorders of Sheep and Goats.* Wolfe Publishing, 1993.
    An indispensable reference for both experienced and student veterinarians, breeders, and others concerned with diseases of sheep and goats worldwide.
Gates, Norman, D.V.M., M.P.H. *A Practical Guide to Sheep Disease Management.* News-Review Publishing Co., 1994.
    The most widely used veterinary reference ever written for the U.S. sheep industry. Every serious sheep producer should own this book.
Pipestone Lamb and Wool Program, Southwestern Technical College, Pipestone, MN. *Pipestone Sheep Management Wheel.*
    Twenty basic sheep management programs outlined.

# 7. Getting a Good Lamb Crop

*"I'll see J. Paul down in a jug with his feet propped up—and he's got the lamb and gets it started nursing."*—DEBBIE BROWN

*"Murphy's Law is when you don't check, something goes wrong."*—TOM CLAYMAN

Lambing time is the yearly test for the shepherd. Attention to your buildings and equipment, the breeding of your sheep, and the health and nutrition of your ewes helps you produce a good lamb crop. However, if your ewes are not responding well to feed, you can change their rations. If the handling chute you built is big enough for a cow to turn around in, you can rebuild it. But when it comes to lambing, if you are misinformed or careless, you may have a dead lamb and a ewe with a scorecard of zero for the year.

This is not to say that everything is in the hands of the shepherd. Even experienced shepherds chalk up a few losses to "lambitis"—unpredictable and unexplained deaths. There is no way to avoid all losses. However, having the right health program, equipment, feed, housing setup for lambing, and a rigorous schedule for checking your ewes will help to avoid them. And, some courage will certainly help, for decisive action is critical at lambing time. The management of breeding and lambing and care of the newborn lamb are all part of producing a good lamb crop.

## EWE EFFICIENCY

The efficiency of your ewes is measured by the number of lambs they produce per year or the lambing average. The lambing average is the subject of much discussion among shepherds. A lambing average of 175% means that many ewes in the flock had twins or triplets. If these lambs are raised properly, your flock will be profitable. A lambing average of 75% means that there were a lot of deaths or that some ewes didn't have lambs at all.

There are two ways to improve the efficiency of a ewe: either she can have more than one lambing per year or she can have more lambs at a yearly lambing. Increasing the number of times per year that a ewe has lambs is called *accelerated lambing*. Lambing can occur at 6- to 8-month intervals with some regularity by some ewes. Accelerated lambing is not easily accomplished and is a better program for more experienced shepherds. Lambs must be weaned on a strict schedule, and the ewes must be carefully monitored to make sure they have rebred. Those that do not get pregnant right away must be bred later, making lambing schedules complex.

Increasing the number of lambs born in a yearly lambing schedule is the most practical approach for the beginner and the farm-flock owner. Twinning is partly controlled by genetics and can be increased by good selection of breeding stock. Twinning is also increased by successive ovulations, thus, breeding later in the season, after the ewe has ovulated a few times, is advantageous. Certain breeds of sheep, such as the Finnsheep discussed in Chapter 4, have strong genetics for a high rate of multiple births.

## BREEDING SEASON PRACTICES

Planning for a successful lambing season begins with late summer or fall breeding. Breeding must be timed so that lambs are born at the best time for your program. Aside from accelerated lambing programs, there are basically two lambing seasons: early and late.

In early lambing seasons, breeding should start on September 1 so that lambs are born in January, February, and early March. An early lambing season will allow market-weight lambs to be ready in May or June, which coincides with the time when there is no field work and laborers are more available on the farm. The disadvantage of early lambing is that it requires more-expensive grain and supplement feeding because less pasture is used. Also, more extensive shelter may be needed for lambing because of the season. People who raise stock for summer fairs and breeders of purebreds use an early lambing program.

The late lambing season takes place in late March, April, and early

May, which means the breeding season begins around November 1. Lambs on this program are fed on pasture through the summer but may have a short period on grain before fall marketing. The late lambing season works better on farms with more pasture. It can also mean a higher lambing average, since ovulation rates seem to be higher later in the fall, and in early breeding, some embryo losses occur if September and October weather is too hot.

Increasing the productivity of the flock depends on a high conception rate. You can achieve this by ensuring the good nutrition of both your ewes and rams and by making sure that every ewe has conceived, or is "settled." As a further effort in good management, you want to control breeding so your lambing period will be 30–45 days long. If your ewes begin lambing in February and don't stop until April, you will be exhausted and unable to observe them closely. Exposing your ewes to a ram with a vasectomy (a teaser ram) a few weeks before breeding or having the bucks penned next to the ewes may help bring them into heat earlier and as a group.

Ewes are in heat for only 20–42 hours. Ovulation is thought to occur toward the end of that period. If the ewe is not bred during estrus or if she fails to conceive, she will have another estrus in 14–17 days. For most sheep, the estrous season (the season when ewes have "heat" periods) is restricted to about 4 months. The estrous cycle seems to be controlled by both light and temperature. Except for breeds known to have long breeding periods, estrus will begin about when the number of daylight hours drops below 14. Lower temperatures also affect the start of estrus; blackface sheep seem to be most sensitive to this influence.

One mature ram should be able to service 35–50 ewes. A ram lamb, in his first breeding season, should only be bred to 15–25 ewes. It is important to keep the ram relatively cool throughout the summer because overheating can affect fertility. In an early lambing program, shear the ram around July 1, leaving the wool long on the brisket, or chest. Rams can be fed some grain about 2 weeks before breeding to increase their stamina if they are not overweight.

You should put ewes on good pasture 14–17 days before breeding, and you should add 0.5 pound of grain to their diet. This "flushing" of the ewe improves the ovulation rate and the chances of multiple births.

In addition to flushing, you should worm ewes a month before breeding, vaccinate for enzootic abortion, and trim hooves. If your ewes have heavy or particularly dirty wool, you should shear around the vulva at this time also. This is called "tagging" and is shown in Figure 7.1.

**7.1.** Tagging–shearing around the vulva prior to mating. (*Courtesy, Sunbeam/Oster Corporation*)

To ensure that ewes are being bred and to get an accurate idea of when your lambing will start, have the ram "mark" the ewes he breeds: leave the brisket wool on the ram unshorn and spread it with a mixture of grease and paint. When the ram mounts the ewe, the mixture will mark her wool. Use a different color of paint for each ovulatory cycle to determine when ewes have been bred: yellow for the first cycle, red for the second, and black for the third. The ovulation cycle lasts roughly 16 days. If a ewe is marked with a yellow paint during the first 16 days and she is not marked later by another color, you can assume that she has "settled" and that she will be among the first batch of ewes to lamb. Those ewes marked with red will lamb later, and those marked with black later still. If a high proportion of your ewes are marked in the third cycle, your ram may not be fertile. Assume that those ewes not marked by the third cycle will not have lambs; they should be culled before a winter's worth of feed is wasted. You may not want to cull the unmarked ewe lambs, as they often do not breed in their first season.

The mixture of grease and paint needs to be applied to the brisket wool every other day because it wears off quickly. A ram marking harness (Fig. 7.2) is a good piece of equipment to buy if you are serious about

**7.2.**   Ram-marking harness.
(*Courtesy, Pipestone Veterinary Supply*)

keeping tabs on your breeding dates. These harnesses come with grease crayon markers that can be changed. The leather harness, though more costly, works best.

Prelambing procedures are summarized in Table 7.1.

## PREPARATIONS FOR LAMBING

The gestation period is 148 days. You should have everything ready for lambing about 1 week ahead of time, in anticipation of any early births. Even the experienced shepherd who plans on February lambs occasionally ends up with a Christmas Eve lamb because he forgot the night in August when the buck jumped the fence. In addition to those matters already discussed, you should take the precautions listed on the next page.

**Table 7.1.   10 steps to a successful breeding season**

1. Shear rams, deworm, trim feet, and protect from heat stress at least 60 days before breeding season. Including teaser rams.
2. Have a breeding soundness exam completed on rams 30–60 days prior to turn out.
3. Vaccinate ewes with enzootic abortion vaccine and deworm 30 days prior to breeding .
4. Condition score ewes and sort into three groups (fat, moderate, and thin) based on body condition. Thin ewes should be supplemented prior to flushing (approximately 2 weeks prior to flushing). Moderate condition ewes will respond best to flushing.
5. Turn teaser ram in with ewes 14 days before turning out fertile rams. Teaser rams are only effective early or late in the breeding season to stimulate ewes to cycle.
6. Start flushing ewes 14 days prior to breeding and continue for 3 weeks into the breeding season. Ewes can be supplemented by ½ to 1 pound of corn and/or turned onto a fresh pasture with lush regrowth.
7. Use adequate ram power. Recommended ratio for ram lambs is 1:25 ewes and mature rams is 1:50 ewes.
8. Avoid handling or stressing ewes during the breeding season. The first month after fertilization is very critical for embryo survival.
9. When breeding during periods of high temperatures, provide shade and fresh water. Heat stress will increase embryonic death loss in ewes and may cause temporary sterility in rams.
10. Use marking device to monitor breeding activity and determine lambing dates.

Source: Jeff Petersen, Lamb and Wool Instructor, Southwest Technical College, Pipestone, MN.

1. Set up a nursery area with lambing pens (described in Chapter 3) to accommodate about 15% of your flock. The area should be clean, well bedded, well ventilated, and as free from drafts as possible so the lambs will not become chilled. Outlets for protected heat lamps are useful. Water and feed should be accessible to your ewes while they are in pens, and there should be an area to mix small groups of ewes and lambs together after they are out of the pens.

2. Have scissors and strong iodine for treating the navels of newborn lambs.

3. Have an emergency supply of colostrum for weak, chilled, or abandoned lambs. (See Chapter 6, Managing Common Health Problems of Lambs, p. 90, #2.)

4. Buy one of the two available types of lamb nipples. One is designed to fit on a soda bottle, and the other comes with its own bottle. Remember that a calf nipple will not work.

5. Decide on a device for feeding severely weakened lambs. Eighteen inches of surgical tubing fitted to a 60-cubic centimeter syringe will do, as will more sophisticated equipment ordered from supply catalogs.

6. Find out where to get powdered lamb milk replacer. In a small flock you may not need it, but check with your feed store to make sure you can get it if you do.

7. Have a detergent available as a lubricant and have antibiotic uterine boluses on hand in case you have to deliver a lamb. You can order the uterine boluses from a catalog or buy them from your veterinarian.

8. Have clean dry rags in the nursery area for drying off lambs in cold weather.

## SIGNS OF LAMBING

Learn how to tell when a ewe is about to lamb. Until you become experienced, the following observations should be helpful:

1. The udder will enlarge before lambing, and just prior to the lambing all the nipples will become distended. Particularly in ewe lambs and yearlings, these changes may be difficult to see unless the ewes have been sheared around the vulva and udder. A full udder may mean different things, depending upon the age of the ewe. An older ewe that has lambed several times may have a noticeably enlarged udder 2 weeks before lambing. A ewe lamb's udder may not look enlarged until just shortly before lambing.

When your ewes are at a feed bunk, you can examine their udders— you can tell a lot by looking. Grab the ewe and examine the udder more closely to see if it is firm and if there is milk when you strip the teats. Those ewes with milk should be put close to the nursery area. Do not put the ewe in a pen yet because it can be several days before lambing, and the ewe should be moving about during that time.

2. While some ewes drop a lamb immediately after eating, most isolate themselves and stop eating before they lamb. A ewe about to lamb may try to claim a lamb born to another ewe and make the chuckling noises of a new mother. She may paw the ground to make a nest, lie down, get up, and paw again. She may wander about nervously, as though she were looking for an already born lamb. Because of the occasionally erratic behavior of the first-time mother, be careful that the ewe lamb isn't looking for a lamb she actually did have.

3. Some bloody or clear discharge often occurs before labor even starts. Labor may not start for another day, but these ewes should go inside.

4. A dark bubble may protrude from the vagina. This is the as yet unbroken amniotic sac, or waterbag. This bubble looks like an emergency, but it is not. You should wait for active labor to begin before you interfere.

5. The area around the vulva becomes swollen and enlarged, another sign that cannot be easily detected unless the ewes have been sheared in this area.

6. Vaginal prolapses are not uncommon toward the end of pregnancy. With just a minor prolapse that moves in and out, the ewe may get through labor without extra help. With a serious prolapse, you will probably have to take the corrective steps described in Chapter 6.

## DELIVERY OF A LAMB

Along with other misinformation about sheep, you may hear that sheep can't give birth unaided. A healthy, properly fed ewe, especially one that has lambed before, will probably deliver her lamb unassisted and take care of it on her own. The less interference the better. Intervention can create its own problems, such as infection from unsanitary delivery methods. You should watch closely and give nature every chance before attempting to intervene. With experience, you will know more about when your assistance is needed. Assuming you do not have that experience, there are two general rules to follow. First, from the time a ewe is in active labor—when she beings to push—give her anywhere from 1 to 2 hours before you try to help. Some experienced shepherds might even give her as long as 4 hours. Second, from the time you see a part of the lamb—a head, a leg, or even a large bulge where the head is pushing out—give the ewe 20-30 minutes before you interfere. Both these rules depend on knowing when the ewe actually went into labor, which means that some guesswork is usually involved.

Once you have decided that the ewe needs help, follow these guidelines:

1. If it is very cold and your ewe is outside, try to get her to shelter, both for your sake and the sake of the new lamb.

2. Wash your hands up to the elbows and use a detergent as a lubricant. A pair of disposable surgical gloves may also be useful.

3. Place your ewe on her side. It is helpful, though not absolutely necessary, to have an extra person to help in case she struggles.

4. Be gentle and take your time. Being gentle is essential because the uterus can be torn with rough handling; a rupture might occur in which even the intestines can spill out. Remember that in most cases the lamb is still safely attached to the mother's supply of oxygen. You may work with the ewe for as long as an hour and still have a perfectly healthy lamb.

5. Examine inside the ewe carefully. The lambs inside the birth canal will feel like a mass of rough, wet, hard spots. Figures 7.3–7.7 show the normal position of the lamb and some of the more common abnormal positions.

The normal position of the lamb (Fig. 7.3) is with its head between its two front legs. If the lamb is in this position but the ewe has been in labor for an extended period of time, the lamb probably has an especially large head. You will need to work your hands around the head and attempt to ease it through the opening. If you can, pull the legs out first. The legs should be pulled out and down toward the udder, not straight out. Sometimes the lamb's head is turned backward into the womb (Fig. 7.4). You

**7.3.** Normal presentation of the lamb. (*Courtesy, Iowa State University Cooperative Extension Service*)

**7.4.** Presentation with head turned back. (*Courtesy, Iowa State University Cooperative Extension Service*)

**7.5.** Presentation with front legs back. (*Courtesy, Iowa State University Cooperative Extension Service*)

**7.6.** Breech presentation. (*Courtesy, Iowa State University Cooperative Extension Service*)

**7.7.** Multiple birth. (*Courtesy, Iowa State University Cooperative Extension Service*)

should try gently to manipulate the head around so it is in position for a normal deliver. Figure 7.5 shows a lamb head first but with its feet back. To make this a normal delivery, you will need to push the head back in gently and bring out one leg at a time. Then pull out and down. Figure 7.6 shows a lamb in breeched position; you will feel the tail when you reach inside the ewe. Push the lamb back inside far enough so that you can pull the hind legs out, one at a time. Then pull out and down. In problem cases of multiple births (Figure 7.7), you may find a mass of heads and legs when you reach in. Again, take your time. Sort out a head and front legs that belong to one lamb and attempt to deliver that lamb first.

6. If the uterus comes out after the delivery of the lamb, wash it carefully and put it back in place. If a veterinarian is available, ask him or her to help you with this procedure (see Chapter 6).

7. If you deliver a lamb and it appears to be dead, tickle the inside of its nose with a straw or pinch your fingers around the umbilical cord and slide them down. Some people even try mouth-to-mouth resuscitation. A "lifeless" lamb may give a hearty flop, and the mother may get up to lick it just as if she had given birth unassisted. Examine the ewe again quickly to see if there are any other lambs inside. Usually if the first delivery is a large lamb, the second will follow without difficulty.

8. Use antibiotic uterine boluses after any intervention to prevent infection.

After delivery, follow these guidelines:

1. Place the lamb and its mother in a lambing pen. Most ewes will follow you if you drag the newborn lamb in front of them. However, some

ewes become confused and panicky, especially first-time mothers. Be patient. Keep the lamb in front of the mother, and if you must, leave the lamb briefly to give the mother a chance to return and find it. Keep track of the ewe by checking her ear tag number, for she may actually walk off and forget the lamb completely.

2. Once the ewe and lamb are penned, clip the lamb's navel to about 1 inch and dip the stub in iodine to prevent infection. Keep strong iodine in a wide-mouthed bottle in your lambing area, and invert it right over the lamb's navel. Some shepherds use iodine solutions that come in spray cans.

3. Check to make sure the ewe has milk. You may get under the ewe or lean her against the wall while you strip her teats. Sometimes a plug in the teats is difficult for a new lamb to remove. If milk comes easily, you can leave the mother and lamb(s) alone for a half hour to an hour in order to give the ewe time to dry the lamb(s) off and start nursing. If you want to make sure a lamb has its first milk, hold the ewe and let the lamb nurse. Do not push the lamb's head upward or it will pull back from the nipple. Instead, push the lamb from the rear, or stroke the top of its tail like its mother does.

To the beginner, new lambs will look weak and deflated. But if the milk is coming and the lamb is standing up, this deflated look will disappear within a few days. You will soon learn to recognize the signs of a lamb that is getting what it wants: the aggressive searching and the rapid wiggle of the tail when it finds the teat and gets milk.

4. Develop a system for identifying which lambs belong with which mothers, and mark the lambs as soon as possible. Ewes should already have ear tags with numbers. You can use wool paint and mark each lamb with the same number as its mother, or you can use small metal ear tags for lambs.

## REVIVING CHILLED AND WEAK LAMBS

A chilled, weak lamb that is not otherwise sick is a good candidate for therapy. The lamb may have wandered away from its mother while she was preoccupied with having a second baby, been weakened by a difficult delivery, or been rejected by its mother. Such a lamb may appear so close to death that an effort to save it will seem futile. These lambs, given food and warmth, may be walking around your kitchen in a matter of hours.

Work quickly. Feed the lamb colostrum if you are uncertain whether it has nursed with its mother yet. If you know the lamb has nursed, you may use lamb milk replacer. If the lamb is too weak to nurse, you need to use a tube for feeding. The tubing, 18 inches of surgical tubing connected to a 60-cubic centimeter syringe, should be coated with water or milk before you put it down the lamb's throat. It is possible for the tube to enter the lung area instead of the stomach; pouring liquid into the lungs can cause pneumonia. If you feel air coming out of the empty tube after it is inserted, remove it and try again. Use 60 cubic centimeters, or about 2 ounces, per feeding. This milk should be warm but not hot. To warm the lamb from the outside, some shepherds dip it briefly in warm water and dry it off quickly in a warm place. You can use a well-insulated heating pad or simply a box next to a heater or wood stove. Shortly after the feeding and warming, the inside of the lamb's mouth should begin to feel warmer. A second feeding, 2 hours after the first, may be necessary. If the mother's milk supply is sufficient and the lamb has not been rejected, it should be returned to its mother as soon as it is strong enough to walk around.

The lamb that will not be placed with a mother should receive 50 ounces of colostrum during the first 24 hours after birth, 8 ounces every 4 hours. Extra large lambs may need more. The benefits of colostrum—providing antibodies and a healthy start to life—are not fully understood, but are well established.

## CARE OF ORPHAN LAMBS

If a ewe dies or has poor milk or insufficient milk for twins, it is possible to transplant, or graft, a lamb to another mother. This takes some shrewd planning because the ewe will accept only those lambs that she identifies by smell as her own. Some shepherds are very successful with grafting and take pride in the fact that they raise no bottle lambs. This is a great savings in both the shepherd's time and in cost of milk replacer.

In a small flock, grafting is more difficult simply because there are fewer potential mothers. It is much easier to trick the ewe immediately following birth than after she has had her lamb for a while. The following are some of the shepherd's tricks to graft lambs:

1. Tie a dog near the pen to bring out protective instincts in the mother and prompt her to accept an odd lamb.
2. Rub the orphan lamb with amniotic fluid from the lamb of the adopting ewe. This is sometimes called slime grafting.
3. Dip the ewe's real lamb and the orphan together in a bucket of warm saltwater to mix their scents.

4. Tie the skin of a ewe's dead lamb onto the orphan you want her to adopt.

5. Place the ewe in a stanchion to permit the lamb to nurse. After a few days, the mother may accept the lamb without the stanchion. A simpler way is to hold the ewe four times a day and allow the lamb to nurse. The ewe will gradually get the message and will at least let the lamb nurse when the shepherd is standing by the pen.

In order to guard against ewes abandoning their young, ewes and their lambs should gradually be mixed in with other ewes and lambs. Mothers with single lambs should be let in the lambing pens for 1–2 days, and those with twins for 2–3 days. As you empty the pens, first put the ewes in with three or four other ewes and their lambs and then with a dozen to make sure that each animal knows who belongs to whom. One hundred ewes together with single lambs or 50 with twins are the maximum number you should allow in the same lot for the first month.

## HEALTH PROBLEMS OF YOUNG LAMBS

Most lamb deaths occur in the first few days of life. Starvation is the primary cause. Colostrum deprivation, *E. coli* scours, and pneumonia are other important problems. These and other lamb health problems are covered in Chapter 6.

### Docking and Castration

There are several ways to dock and castrate lambs. They include the hot chisel method, elastrator bands, emasculators, and knives. For castration, the old method of using a knife and the shepherd's teeth is still used and is claimed by some to be highly sanitary and effective. Each method has it advantages and disadvantages. Sheep producers of equal experience will swear by quite different methods. The important points to keep in mind for the beginner are the following:

1. Whichever method you use, perform the operation while the lambs are small, preferably a few days old. The bigger the lamb, the riskier the procedure.

2. Whichever method you use, vaccinate lambs with tetanus antitoxin at the time of surgery.

3. Watch for open wounds and treat them promptly. Maggots infect wounds that are not properly closed. Maggots are a greater problem in a late lambing program.

4. Castration is not necessary in ram lambs that will be marketed by 5 months of age and before July 4.

5. Perform docking on all lambs. The long woolly tail not only becomes unsightly, but it also makes the lamb vulnerable to maggot infestation.

6. Figures 7.8 through 7.11 demonstrate how to castrate a lamb with a knife and dock a lamb's tail with an elastrator. Whatever method you use, have a veterinarian or experienced shepherd give you a demonstration the first time. Improper docking and castrating may be ineffective and can even cause death.

**7.8.** Assistant holds lamb in this position for docking and castrating.

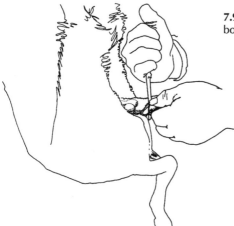

**7.9.** With a sharp knife, cut off the bottom third of the scrotum.

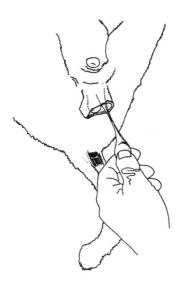

**7.10.** Push the scrotum up against the belly of the lamb so that the testicles are exposed, using your right hand. With the left hand, grab one testicle between thumb and forefinger and pull until the vessels snap. Repeat with the other testicle. The cut should be treated with iodine or antiseptic and in warm weather, the site should be sprayed with a screwworm bomb.

**7.11.** Proper site for docking is a point just below where the ligaments go into the tail. This allows the tail to cover the rectum. The elastrator band is shown here, but the site would be the same regardless of the method used.

## YOUR RESPONSIBILITY

Most lamb deaths occur in the first week or two of life. Many of these deaths can be avoided by proper attention from the shepherd. Watching over the ewes before birth, caring for the newborn lambs and mothers, and providing adequate health protection are the keys to the survival of your lambs.

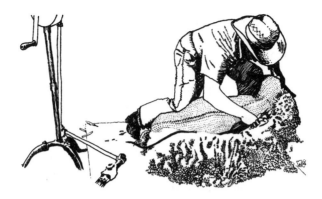

# 8. Wool Production

*"I walk around with a piece of fleece in my jacket pocket at all times. I carry something in a beautiful grey or silver or brownish color. That's marketing."*
—RANDY IRWIN

*"I've had Michael call all around the country to find out what mills were paying for wool."*
—TOM CLAYMAN

Wool is one of the marvels of nature. Through different milling processes, it can be made into fine worsted fabrics such as gabardines and twills, felt, or bulky knits for sweaters. Wool is warm and strong and has good flexibility. It has the capacity to absorb large amounts of moisture, yet in either warm or cool weather it won't feel clammy. It lets air in and out, which also contributes to its comfort. Because it absorbs dyes very well, wool is especially beautiful. Moreover, wool retards fire; although it can burn, it does not flare up or melt on the skin. Washable wools can now be produced, taking away the stigma that wool is expensive to care for.

Wool production in the United States, along with sheep production as a whole, has been on a steady decline since the 1980s. The 1994 wool poundage was the lowest since records have been kept—68 million pounds compared with 143 million in 1973. When compared with other fibers on a per capita basis in the United States, "consumption" of wool is only 1.7%, whereas manufactured fibers are 57%, cotton is 37.5%, and "other" is 3.8%. (1) The ending of the wool incentive payments in 1996 also brings some changes for producers. Some feel that this may actually be good for the industry, by forcing producers to concentrate on quality and not relying on the price supports.

There is a market for wool in this country, and imported raw wool now accounts for 65% of our domestic supply. Domestic mills cannot find enough quality wool to meet their needs. Part of the problem is the quality of the domestic wool, which has been neglected for so long.

In the 1950s and 1960s, income derived from wool was sometimes barely enough to pay the shearer. But despite the big trends, the wool crop, when properly managed, can make a substantial contribution to your sheep operation. The percentage varies from 10% to 50% of income in different parts of the country. The farm-flock states, where the emphasis is on meat-producing sheep, are at the low end. The western states are at the high end. One rule of thumb is that the wool crop will pay for the ewes' feed for the year. To neglect the wool crop is a serious mistake.

Wool processing is complicated, but the basic steps all have their roots in the ancient methods of changing fleece to fiber. The fleeces are first sorted according to their type and quality, for these traits determine how they are used. The wool must then be cleaned to remove natural grease (lanolin) and any dirt or other impurities. Carding, which involves separating the fibers so they will be straight, comes next. Carding produces a web of thin batting, which is spun and woven. The weave is then tightened with heat and moisture by a process known as fulling.

While the technical aspects of this process may not interest you as a sheep producer, you should understand how this process is related to the wool you sell. Wool that is difficult for the mill to process or that is not the right type for current demands will not bring the best price.

"Raw" or "grease" wool are terms used for the fleece after it has been sheared from the sheep. The relative worth of this raw wool depends on three main features: the diameter of the fiber (fineness), the staple of the fiber (length), and the yield (amount of shrinkage).

The diameter of the fiber is the actual thickness of an individual strand of wool. The diameter determines the grade of the wool. The grading of wool was originally based on a system that relied on bloodlines. The Merino sheep was the standard, and the coarser the wool, the less Merino blood the sheep was said to have; for example, one-quarter blood wool was coarser than one-half blood wool.

Today, the most commonly used system of grading is the count system, which is based on the number of hanks of yarn that can be spun from a pound of clean wool. A "hank" is an old term that refers to a length of yarn 560 yards long. The electronic microscope provides an even more accurate measure of the diameter of yarn, but the count system is still widely used.

The wool grades are fine, medium, coarse, and very coarse. Breeds vary in these traits, but individuals of a breed will not necessarily conform to this grade, and all the wool from a single sheep will not provide exactly the same grade. The wool on the shoulder is usually the finest and that on the britch (or breech, the outside of the rear legs) the coarsest. The grade of a fleece refers to the general level of fineness. There is some confusion over the vocabulary and measurements used in this area of assessment. The count traditionally referred to the crimp, the number of waves that the fiber had per unit. The crimp is generally inversely proportional to the diameter, that is the greater the number of crimps, the smaller the diameter of the wool. In subjective assessments of wool, it was the crimp that the buyers were generally looking at to assess diameter. The United States is moving toward a system that measures the diameter of the fiber by scientific means.

Staple length refers to the length of a small swatch of wool fibers. Wool is generally divided into two staple classes: apparel wool and carpet wool. Apparel wool is short and fine, carpet wool is long and coarse. The coarser the wool, the longer it must be before it can be combed to make the fibers lie parallel. In the United States, most of the wool sold is apparel wool. Coarser grades are imported from New Zealand.

The third criterion for judging and pricing a fleece is the yield, also commonly called shrink. This is the percentage of clean wool ready for processing after grease, burrs, dirt, and other foreign material are removed. A 10-pound fleece may be reduced to 4 1/2 pounds of clean wool; in that case, the yield would be 45%, and the amount of shrink is 55%. The yield varies with different grades. Generally, the finer the wool the lower the yield. The estimated shrinkage of the fine wool produced in the western states can be as high as 75%. Table 8.1 shows the fleece yield and fiber diameter of wool from different breeds of sheep.

## CARE OF THE WOOL CROP

The three main areas of concern—grade, staple, and yield—can all be considered in your production plan. The breeds raised will determine the grade and class of wool of a flock. Above and beyond the variations in shrink based on grade, yields of wool can be improved by good care.

What can be done to improve the wool crop? The price you receive for your wool is docked, or lowered, if it requires extra processing to clean. The price is also docked if wool is in otherwise poor condition. For example, it may have too many second cuts, or short fibers produced when the shearer goes over the sheep twice. Here are some simple rules that will prevent unnecessary docking when you market your wool.

**TABLE 8.1.   Fiber diameter and fleece weight and yield by breed**

| Breed | Average fiber diameter (microns) | Range of weight of fleece (lb) | Range of yield (%) |
|---|---|---|---|
| Border Leicester | 38–30 | 8–12 | 60–70 |
| Cheviot | 33–27 | 5–8 | 50–65 |
| Columbia | 30–23 | 9–14 | 45–55 |
| Corriedale | 31–24 | 9–14 | 45–55 |
| Delaine–Merino | 22–17 | 9–14 | 40–50 |
| Dorset | 33–27 | 5–8 | 50–65 |
| Finnsheep | 31–24 | 4–8 | 50–70 |
| Hampshire | 33–25 | 5–8 | 50–60 |
| Lincoln | 41–34 | 10–14 | 55–70 |
| Montadale | 30–25 | 5–9 | 45–60 |
| Oxford | 34–30 | 5–8 | 50–60 |
| Rambouillet | 23–19 | 9–14 | 45–55 |
| Romney | 39–32 | 8–12 | 55–70 |
| Shropshire | 33–25 | 5–8 | 50–60 |
| Southdown | 29–24 | 5–8 | 40–55 |
| Suffolk | 33–26 | 4–8 | 50–60 |
| Targhee | 25–21 | 9–14 | 45–55 |

Source: From *Sheep Production Handbook.* American Sheep Industry Association, Englewood, Colorado, 1988. (With permission)

1. Feed sheep adequately, using the recommendations in Chapter 5 as a guide. Scrawny, underfed sheep will produce both poor lamb and wool crops.

2. Use well-designed hay feeders and grain feeders. Feeders of the wrong size or design permit sheep to rub grain and chaff into the wool.

3. Watch your pasture for burrs and get rid of them.

4. Do not brand your sheep with anything but acceptable livestock marking crayons or wool paint, both of which will wash out.

5. Try to keep your sheep free of dung locks, especially the longer-wooled sheep. The whiter the fleece, the more valuable it is. Stains from dung locks will lower the price of your wool.

6. Employ a good shearer, one who does not make second cuts.

7. Bag separately tags and those swatches of wool that are clumped with manure and damp with water or urine.

8. Do not use anything except special paper twine to tie fleece. Wire and baling twine, especially the black plastic kind, present problems in handling.

9. Do not use a fleece-tying box (a specially hinged wooden box that allows you to tie the wool into a neat, easy-to-manage package). Tying boxes use more twine, can pack the fleeces too tightly, and cause the wool to mildew.

10. Tie your fleeces properly. Put the clean part of the fleece (the part that has been next to the skin) on the outside. That way the buyer can judge its quality without untying the bundle or pulling the fleece apart.

11. Bag the wool in new burlap wool bags, not in paper or plastic bags.

## PROCEDURE FOR SHEARING

Shearing traditionally takes place in March or April, before sheep are turned out to pasture. In recent years, increasing numbers of shepherds shear their sheep in the winter, a few weeks before lambing, thus, the mothers are cleaner and less apt to wander out in a storm to lamb; however, winter shearing requires ample indoor housing. Still other shepherds shear twice a year, once before lambing and once in midsummer. In the twice-a-year program, you may get a pound more wool per ewe for that year, but the wool will be shorter and might bring a lower price. Lambs being fed for market during the summer in a late lambing program are usually sheared in July, because it is thought to improve weight gain in hot weather.

Shearing is an ancient process and requires a fascinating combination of skill and art. For most of its early history, shearing was done with hand clippers like those shown in Figure 8.1. Today, modern shearing equipment is generally used. Figure 8.2 shows a shearer using small, electric hand clippers, which are useful for the small-flock owner. Professional shearers use more-powerful machines.

There are three main methods of shearing used in this country: the

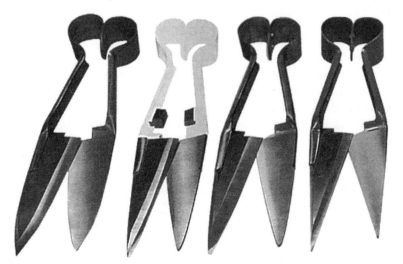

**8.1.** Hand clippers. (*Courtesy, Premier Sheep Supplies*)

**8.2.** Modern electric hand clippers. (*Courtesy, Premier Sheep Supplies*)

Australian, the American open, and the Mexican. People who observe an experienced shearer at work often think the sheep has been especially trained for the performance because it does not struggle. The shearer holds the sheep in special positions to keep the skin tight, avoid cuts, and make the sheep comfortable and unable to struggle.

Many state extension services or other schools organize shearing clinics, where you can learn how to shear. You can find out when a clinic will be scheduled in your area by writing to the sheep specialist at the extension service located at the land-grant university in your state. These offices are listed in Appendix A. With the declining number of shearers, it may be worth your time to learn this challenging skill. Many shearers learn by apprenticing themselves to an experienced shepherd or to a professional shearer.

## THE SHEEP SHEARER

The numbers of shearers vary greatly from one part of the country to the next; in some areas you have a choice, in others you may have to wait months for the only local shearer to arrive at your farm. Your extension service can probably supply you with the names of shearers; check the reputation of these shearers with local sheep producers. Normally, shearers charge a flat fee for each sheep, but some also charge a setup fee for shepherds with small flocks. Here are some guidelines you should use in dealing with shearers.

1. When you call to make arrangements, get both a date and a price for the shearing.

2. Be certain that your sheep are dry when they are sheared. If the shearer is coming in the morning, consider penning the sheep inside the night before so that the morning's dew will not dampen the wool. If there is a chance of rain, pen the sheep inside. Damp sheep will dry off fairly quickly if they are penned inside.

3. Do not crowd the sheep, but be sure they are penned in an area easily accessible to the shearer, who will not want to chase sheep or wait while you set up a pen.

4. Have a clean, unbedded, level floor on which the shearer can work.

5. If the shearer does not have a paid assistant to catch sheep and tie wool, offer to help with these jobs.

6. Ask the sheep shearer questions, for they often raise sheep themselves and will probably be a good source of information.

Some beginners panic when they see cuts and scrapes on their newly sheared sheep. Small cuts are part of the shearing process, and the lanolin secreted by sheep has strong healing properties. Larger cuts may need to be sprayed with iodine or treated with pine tar to keep the flies away. Accidents can happen, but serious wounds are signs of a careless shearer. A large number of second cuts are also evidence of a poor shearing job. The shearer should end up with the fleece in one piece so that it can be tied properly (Fig. 8.3). Lamb's wool is too short to be tied and should be placed in a bag.

## MARKETING WOOL

The system of marketing wool is usually difficult for beginners to understand because of the many numbers and unfamiliar vocabulary. Most beginners, and even many experienced sheep producers, tend to sell to a shearer or a wool buyer for whatever price they offer. However, increasingly producers are paying more attention to the care and marketing of the wool crop.

Serious producers are "skirting" their wool before trying to market it. Skirting is a process whereby the wool is laid out flat and britches and belly wool, which bring a lower price, are taken out. This is proving to be a way to get a better price for the wool.

Your extension service can supply you with the names of wool buyers. Some areas also have wool pools where wool is sold collectively. Before you sell your wool, evaluate it, using some of the criteria discussed in this chapter, and get a price range from several different people. If you have a lot of wool, or are pooling your wool with other people, you may want to have your wool "cored." This is a process of sampling the wool, doing objective tests of the grade, staple, and yield so that you are marketing on the basis of solid measures instead of the subjective assessment of the buyer. If you feel your wool meets high standards and you are not being offered a top price by one buyer, seek estimates from other buyers. Per-

**8.3.** The amazing product of a good shear—the fleece in one piece. (*Courtesy, Premier Sheep Supplies*)

haps you are missing some legitimate shortcoming, or perhaps the buyer is not offering you the best possible price.

## THE PELT

The pelt is the skin and wool removed from a slaughtered sheep. A "number 1" lamb pelt is the best to bring to market. A number 1 pelt is defined as one with wool at least 1/2 inch long, or about 7–8 weeks of growth following shearing.

The market price for lambs with a number 1 pelt might be $2–3 higher per 100 pounds than for lambs with full wool because of the higher dressing percentage, or because of the weight of the carcass ready for cutting as compared with the weight of the lamb when alive. The lamb with the number 1 pelt may also have a 52% dressing percentage, whereas a fully wooled lamb might yield only 47–48%. Also, when packers sell loads of lamb pelts to a tannery, they get the best price for number 1 pelts. The packer should, therefore, give a higher price for number 1-pelted animals.

The freshly sheared sheep will not make a very attractive pelt, but pelts at all other stages can be used in the home for rugs or mittens, hats, or vests. Some producers have pelts tanned commercially and sell them at fairs and shows for a profit. Since having pelts tanned professionally is fairly expensive, consider doing the job yourself at least once. Washing a dirty gray pelt in detergent and water and watching the beautiful transformation can be very satisfying. Notice that the following recipe for tanning does not require the use of caustic chemicals, so even children can help in the process:

## Tanning mixture for 220 pounds of wet pelts
**5.5 pounds of alum**
**2 pounds of salt**
**11 pounds of flour**
**4 ounces of olive oil**
**50 egg yolks**

First, carefully scrape the pelt free of any remnants of meat or fat. Next, wash and rinse it in detergent and water—use a washing machine if you wish. Make a paste of the flour and water; set aside. Mix the egg yolks with a little water and add the olive oil. Blend this mixture into the flour and water paste. Mix salt and alum, and add this mixture to the paste. Spread the paste liberally over the pelt; fold it over, skin sides together, and let it dry for 4–5 days. Wash off the hardened paste with warm water and let the skin absorb the water. As it dries again, pull it and work it to make it soft. This is the difficult part of the process, but the results are well worth the effort.

## WOOL FOR SPINNING

The market for high-quality and unusual fleeces for spinners and other crafts people has been growing. The standards for this market are not the same as for the mills, but it is a market that may well be worth getting to know, for prices per pound are considerably higher if the wool is properly cared for. Call your extension service to find out about shows and fairs where wool for this market is the focus.

Wool has been a neglected part of sheep production in many parts of the country. Making the most of this product should be a goal of your operation, regardless of its size. If you value the wool product from your own flock, you are also apt to value useful things made from wool, whether bought or made at home. And, it is important in today's market that you advertise this value through your own use.

## REFERENCE

1. American Sheep Industry Association. U.S. Sheep Industry Market Information Report. Englewood, CO: ASIA, 1995.

# 9. Management

*"The year we started recordkeeping was the year we made the big jump in our crop."*—DEBBIE BROWN

*"It was a management thing. We needed to get serious and do this the right way, or not do it at all."*
—BILL AND JAN BUTLER

Good management makes the difference between profits and losses in sheep operations. Good shepherds recognize this, for they know how well the time and effort pays off. And the sheep benefit from good management, because well-planned care and feeding makes them comfortable and productive.

Good management means developing a plan that includes marketing, genetics, nutrition, equipment, health care, reproduction, and lambing. Marketing is listed first because what you produce and how and when you sell that product will influence all your other decisions. Marketing plans will differ, of course, depending on the operation, but the key to a successful sheep operation is a clear, comprehensive, well-thought-out plan. In developing and evaluating your plan, consider the following four basic steps:

1. Identify your goals. Perhaps in your first lambing period you had only a 118% lamb crop and you want to improve that figure.
2. Gather facts. First, look at your own records for the lambing period. Was there a high rate of multiple births, but many of your lambs died early? If so, were these deaths due to cold weather or disease? Could

more vigilance on your part have decreased the losses? Did many of your ewes fail to produce lambs? If so, should you consider the ram to be at fault? Or did most of your ewes have single births? Would flushing or later breeding improve your lambing average?

3. Make decisions and take action—try different approaches. For example, when you look at your records you may find that 5 ewes out of your flock of 40 did not have lambs at all. The ewes that did have lambs produced a 150% lamb crop, but 6 of those lambs died in the first day or two. Perhaps some of these lambs died in the cold because they were not brought inside quickly enough due to an extended lambing period. In your next lambing season, you should cull and replace the barren ewes. You also want to better control your fall breeding so your ewes lamb close to the same time, making it possible for you to check the ewes through the lambing season without extreme fatigue.

4. Observe, keep good records, and evaluate results. Suppose that due to your increased vigilance and improved breeding stock, during your second lambing season all your ewes have lambs and you have only two losses—a 145% lamb crop, which is respectable. Or, even if there was no improvement, it could be the result of a whole new set of problems that year. The shepherd must always gather facts, go back and see what worked and what didn't work, and identify new goals. Recordkeeping is essential. Only through proper recordkeeping can you make evaluations that lead to improvement. The larger the flock and the more serious you are about making it profitable, the more important good recordkeeping is. Good management will include a system of recordkeeping that is in line with the goals of the operation and the commitment of the shepherd to maintaining them.

The first step toward good management is to understand the year's calendar for raising sheep. A general calendar is provided with a checklist of necessary activities for each month. Both an early and a late lambing program are included in the calendar. The calendar should be altered to meet the particular needs of your operation.

# SHEEP YEAR

This calendar will help you direct your sheep operation and will put the subjects covered in the various chapters into a working perspective. The calendar is meant to be a useful tool for you to use at your desk or in your barn. You can adapt it to suit your specific situations and recommendations of your veterinarian. At the end of the calendar is a sample barn sheet and a ewe record, which, if enlarged, will serve as a basic recordkeeping device in your lambing area. These records can then be transferred to more permanent records, either on paper or into a computer, when time permits.

The suggestions under each month's calendar are divided into three sections. The general procedures apply to your flock, regardless of the lambing schedule. Early lambing applies to a February or early March lambing program. Late lambing applies to an April or early May program. In both cases, the breeding period is limited to from 35 to 45 days. You may be using both early and late programs with separate groups of ewes. (The calendar is adapted largely from the Sheep Flock Management Calendar prepared by Don Morrical, sheep specialist with the Iowa State University Cooperative Extension Service.)

# JANUARY

### General Procedures

Provide clean, ice-free water; plain salt; and a mixture of limestone, dicalcium phosphate, and trace mineral salt.

Increase feed in extremely cold weather (-5° F).

Worm ewes (if you did not do so in November), using approved sheep wormers.

Examine ewes for ticks and lice. Treat if infected.

Check ewes daily for health problems such as lambing paralysis, footrot, abortion, and respiratory infection.

Vaccinate ewes against overeating and tetanus.

If lamb pneumonia has been a problem in the past, consider adding sulfadimethoxine or tetracycline to drinking water.

### Early Lambing

Gradually increase ewes' feed to 4.0–4.5 pounds of hay, 1–2.5 pounds of grain or equivalent late-gestation ration (discussed in Chapter 6) for the average ewe.

Make sure the lambing area and supplies are ready.

Have a method for identifying ewes and their lambs.

Watch for signs of lambing. Check udders of ewes approaching lambing time as described in Chapter 8.

Identify ewes dropping twins; early-born twin ewe lambs may make good replacement ewes.

### Late Lambing

Market ewes that have not been bred.

Feed 3.5–4 pounds of hay per day or equivalent early-gestation feed.

# FEBRUARY

## General Procedures

Provide clean, ice-free water; plain salt; and mixture of limestone, dicalcium phosphate, and trace mineral salt.

Check ewes daily for signs of health problems.

## Early Lambing

Keep lambing area clean, dry, and well bedded.

Strip teats of ewes after lambing and treat navels of lambs with iodine.

Make sure all new lambs have nursed within 1 hour after birth.

Identify ewes and their lambs with tags or paint brands.

Have a plan to handle orphan lambs. Have lamb milk replacer.

Dock lambs at 2–3 days. Cut tail so anus is covered (see Chapter 8).

Castrate all lambs except those to be retained as breeding rams (see Chapter 8).

Vaccinate lambs for overeating and tetanus at 21–35 days.

Watch for mastitis in ewes. Watch for humpback, stiffness, or scours in lambs.

Feed ewes 1.5–2 pounds grain and 4–6 pounds hay (adjusting upward for ewes with twins, downward for ewes with singles) or other early lactation feed plan.

Pen ewes with single lambs for 1–2 days, and ewes with twins for 2–3 days in lambing jugs.

Start lambs on creep feed in well-bedded creep area.

## Late Lambing

Begin month with ewes on early gestation feed; change to late gestation feed gradually, depending on expected lambing date.

# MARCH

### General Procedures

Provide clean, ice-free water; plain salt; and a mixture of limestone, dicalcium phosphate, and trace mineral salt, free choice.

Check ewes daily for signs of health problems.

### Early Lambing

Shear flock, if you are not on January shearing program.

Treat lambs and ewes for ticks and lice (if present) 4–7 days after shearing.

Continue to feed ewes for maximum milk production.

Continue to creep feed lambs.

Identify ewes that should be culled, based on age, lambing record, or health problems.

Vaccinate lambs for overeating, two injections 10–15 days apart.

### Late Lambing

Watch for signs of lambing paralysis in ewes.

Start March 1 to gradually increase from early to late gestation feed, such as 4.5–5 pounds of hay and .75–1.25 pounds of grain.

Prepare lambing facilities and equipment if shed lambing.

Begin daily inspections for signs of lambing.

# APRIL

## General Procedures

Provide clean water; plain salt; and a mixture of limestone, dicalcium phosphate, and trace mineral salt, free choice.

Check ewes daily for signs of health problems.

## Early Lambing

Reduce the amount of grain fed to ewes after 6-7 weeks of nursing.

Condition ewes for weaning. Reduce feed to maintenance level for 1 week prior to weaning and remove water for 24-36 hours at weaning.

Wean lambs, leaving lambs in familiar surroundings.

Treat lambs for internal parasites before turning out to pasture.

Turn ewes out to pasture when vegetation is 4-5 inches high.

## Late Lambing

Keep lambing quarters clean.

Strip teats of ewes, and treat navels of newborn lambs right away.

Make sure lambs have nursed soon after birth.

Confine ewes with single lambs for 1-2 days, twins for 2-3 days, in lambing pens.

Mark ewes and lambs with tags or paint. Identify possible replacement ewe lambs.

Group ewes and lambs together gradually.

Feed ewes 1.25-2 pounds of grain and free feed hay, which means to give them hay so they can eat what they want, or equivalent early-lactation feed.

Try to feed ewes with twins and ewes with singles separately, giving the latter somewhat less feed.

Dock all lambs and castrate males at 1-4 days of age; vaccinate for tetanus.

Watch for mastitis in ewes and stiffness or scours in lambs.

# MAY

### General Procedures
Provide clean water, fresh minerals, and salt to ewes on pasture.

Check lambs and ewes daily for signs of health problems.

Watch for footrot; treat at first signs.

### Early Lambing
Feed lambs growing and finishing ration.

Feed replacement ewe lambs growing ration.

Market cull ewes for slaughter.

Weigh lambs when oldest is 120 days, keeping records of these weights. Select oldest and heaviest ewe lambs for replacement twins if possible.

Separate replacement ewe lambs from market lambs.

Get quotations on spring lamb prices from markets.

Market groups of lambs as they reach 120–130 pounds or current market standard.

### Late Lambing
Shear ewes and treat ewes and lambs for ticks and lice (if present) 7–14 days after shearing.

Worm ewes not wormed recently.

Fix cutting chute so lambs can be kept in the lot if coyotes or dogs are a problem during daytime grazing.

Dock and castrate lambs.

Identify cull ewes.

Vaccinate lambs for overeating and tetanus by giving two injections 10–15 days apart.

# JUNE

## General Procedures

Provide clean water, salt, and minerals.

Check ewes and lambs daily for health problems.

Watch for maggot infestations, especially around docks; treat accordingly.

Repair prolapsed lambs promptly and market them.

## Early Lambing

Observe lambs for coccidiosis. When lambs die, take steps to find out why, especially if there is more than one death. Don't treat blindly.

Treat unthrifty market lambs and replacement ewes for worms.

Make final decision on replacement ewe lambs.

Feed growing/finishing ration to lambs and growing ration to replacements.

Sheer lightest market lambs (under 75 pounds) late in the month.

Manage pasture to maximize returns: rotate, adjust, stock, use electric fence where helpful.

## Late Lambing

Evaluate worm infestation of lambs. Treat, using approved wormers, or treat routinely at the beginning of every month.

Move ewes and lambs to second pasture. Add supplemental feed if lambs are not growing well. Stock pastures to maximize milk production. Use electric fence where feasible.

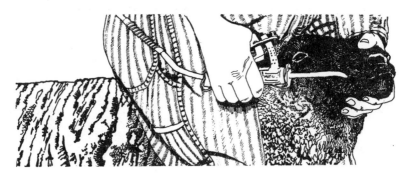

# JULY

### General Procedures

Provide fresh water, salt, and minerals.

Observe for footrot, maggots, and unthriftiness due to worm infestation.

Check ewes for bad udders and footrot. Cull and market old, unsound, and unproductive ewes.

Buy new rams, if necessary, free of footrot, ticks, and epididymitis.

Shear rams and replacement ewes early in the month.

### Early Lambing

Feed replacement concentrate to ewe lambs if out to pasture, or growing ration if in confinement.

Monitor early lambs for condition and sell when correctly finished.

### Late Lambing

Wean lambs during the month. Weaning may be delayed if lambs are under 90 days and pasture is good. Drylot weaned lambs.

Postweaning, shear lambs.

Start weaned lambs gradually on 15–16% protein, 65% total digestible nutrient growing/finishing ration.

Provide shade and breezy area for drylot lambs.

# AUGUST

## General Procedures

Provide fresh water, salt, and minerals.

Treat all ewes and rams for internal parasites (worms) during the second week of the month.

Examine for footrot; trim feet as needed.

Keep all newly purchased ewes and rams isolated for 4 weeks to check for health problems. Vaccinate new ewes for abortion. Tag replacement ewes for record system.

Plan for adequate ram power for your flock, 1 ram lamb per 25 ewes.

## Early Lambing

Vaccinate ewes for abortion 2 weeks before breeding.

Condition score ewes. If some are too fat, drylot them around August 1 and feed below maintenance level until August 20.

Rotate ewes to good pasture on August 20; begin adding flushing feed.

Condition score ewes and put excessively thin ewes on better feed.

Condition rams for breeding, and provide shade for them.

If possible, put rams in close proximity to ewes around August 20.

Conduct breeding soundness exam on breeding rams.

## Late Lambing

Observe lambs for coccidiosis, rectal prolapse, or enterotoxemia.

Keep treating unthrifty lambs for worms.

Feed growing/finishing ration to lambs.

Feed ewe lamb replacements for normal growth.

Market groups of lambs as they reach 120 to 130 pounds or current market standard.

# SEPTEMBER

### General Procedures
Provide fresh water, salt, and minerals.

Observe rams for lameness, swollen testicles, or anything that would hamper breeding.

### Early Lambing
Provide pasture during breeding season.

Stop flushing 3 weeks into breeding.

Continue feeding grain to rams during breeding season.

Turn rams in with ewes about September 1.

Use marking harness for rams, changing color every 15–16 days.

Keep track of ewes bred by color markings.

If you have more than one ram, rotate rams, letting each ram rest 1 out of 3 days.

### Late Lambing
Continue to observe ewes and treat for worms.

Reduce diet roughage to 15–20% in market lambs.

Continue to market 120- to 130-pound lambs. Try to have all lambs sold by January 1.

# OCTOBER

## General Procedures

Provide clean water, iodized salt, and trace mineral salt.

Vaccinate ewes for enterotoxemia 2–3 weeks before turning them out to harvested fields.

Provide supplemental protein and minerals for ewes that are gleaning cornfields.

Put ewes on maintenance ration or allow them to gain slightly.

## Early Lambing

Complete breeding season by October 15, and separate rams from ewes.

Breed ewe lambs separately.

## Late Lambing

Treat all ewes and rams for worms in mid-October.

Drylot very thin ewes October 1; feed above maintenance level until October 20.

Put ewes on good pasture or higher nutritional level for flushing 3 weeks before breeding.

Condition rams for breeding season.

Vaccinate for vibrio and chlamydia before breeding. Give second shot in mid-January if this is a first vaccination, not a booster.

# NOVEMBER

### General Procedures

Provide fresh water, iodized salt, and trace mineral salt.

Treat for worms when you drylot sheep for winter or when they are at midgestation. Use a product that is safe for pregnant ewes.

Check ewes daily for any health problems.

Feed about 2.5–4.5 pounds of good-quality hay per ewe daily, or its equivalent for average-sized ewes.

Repair equipment, feed bunks, lambing panels, and lot fences.

### Early Lambing

Pregnancy check all open ewes; ex-pose them to a ram for possible late lambs.

Vaccinate a second time for vibrio, if ewes have not been treated in previous years.

### Late Lambing

Turn rams in with ewes, depending on target lambing date; use marking harness or grease-painted brisket wool.

Start system of recording when ewes are bred. Change color on marking harness or wool every 15–16 days.

Provide good pasture or equivalent in drylot feed.

Observe breeding behavior of rams.

# DECEMBER

## General Procedures

Provide fresh, ice-free water; plain salt; and mineral mixture of limestone, dicalcium phosphate (including selenium), and trace mineral salt, free choice.

Plan health program for the year.

Provide early-gestation ration for all ewes.

Feed bred ewe lambs separately from mature ewes.

Plan lambing area and system for handling lambing time.

Use reduced labor period to attend meetings related to raising sheep, read about sheep, and visit other sheep producers.

Consider giving wool and lamb products for holiday gifts.

## Early Lambing

Treat for internal parasites, using medicines, such as Tramisol or other approved wormers, that are safe for pregnant ewes.

Late in month, change to late-gestation ration.

Market open ewe lambs at choice grade.

## Late Lambing

Complete breeding season 35–45 days after rams were turned in.

This calendar makes clear the importance of records. A barn sheet (Fig. 9.1) and ewe card (Fig. 9.2) are two examples of record forms that are important for almost every shepherd, even those with small flocks. Records of slaughter lambs, numbers of ewes, amount of wool sold, and amount of feed used are also essential. These forms may be available from your extension service. The serious sheep producer will use records to evaluate the overall profitability of his or her operation.

A variety of computerized recordkeeping supplies are available for purchase. It is also possible to build a system to suit your own needs using a spreadsheet. The important thing is to have a system that tracks the most important part of your operation, and use it faithfully.

As a beginner, managing a flock may seem overwhelming. At times the problems may come faster than you can handle them. During your first lambing, when you are weary and confused and chilled lambs and ewes with exotic health problems seem to be everywhere, remember that these experiences make you a better manager. You need to record, evaluate, and transform these experiences into a better plan for the future.

**Barn Sheet**
For Ewe Records    Year _____ Page _____

| Ewe Number | Lambing Date | No. Born | Service Sire | Lamb Number | Sex | Birth Wt. | Weaning Wt. | Date Weaned | Adjusted Wn. Wt. | Remarks | Fleece Weight |
|---|---|---|---|---|---|---|---|---|---|---|---|
| | | | | | | | | | | | |
| | | | | | | | | | | | |
| | | | | | | | | | | | |
| | | | | | | | | | | | |
| | | | | | | | | | | | |
| | | | | | | | | | | | |
| | | | | | | | | | | | |
| | | | | | | | | | | | |
| | | | | | | | | | | | |
| | | | | | | | | | | | |
| | | | | | | | | | | | |
| | | | | | | | | | | | |
| | | | | | | | | | | | |
| | | | | | | | | | | | |
| | | | | | | | | | | | |
| | | | | | | | | | | | |
| | | | | | | | | | | | |
| | | | | | | | | | | | |
| | | | | | | | | | | | |

**9.1.** A barn sheet for ewe records. (*Courtesy, Cooperative Extension Service, Iowa State University and the USDA*)

## Permanent Ewe Record Card

| | | Notes: |
|---|---|---|
| | Sire No. | |
| Ewe No. | | |
| | Dam No. | |
| Birthdate | S-Tw-Tr | |
| Adj. wt. | wt. ratio | AS481(b) |

Ewe No. _____

| YEAR | Lamb No's. | Adj. 120 wt. | Wool wt. x 2 | Index or Ratio |
|---|---|---|---|---|
| | | | | |
| | | | | |
| | | | | |
| | | | | |
| | | | | |
| | | | | |
| | | | | |
| | | | | |
| | | | | |

**9.2.** Both sides of a permanent ewe record card. (*Courtesy, Cooperative Extension Service, Iowa State University and the USDA*)

# 10. Cooking with Lamb

*"I don't understand the problem with the United States and lamb. It's the preferred meat of the world."*—JEAN BROWN

*"Our focus has been 'Do it quick, do it simple.' Then we add 'It's different. Take a risk!'"*
—SPENCE RULE

It is both the natural inclination and the responsibility of the shepherd to promote lamb, and one good way to do so is to serve lamb often and in attractive and tasty ways. For many beginning shepherds, lamb will be a new food. Not only is lamb consumed in limited quantities (0.9 pounds per capita in 1995) in this country, but it is a new dish for most American cooks unless they are from cultural backgrounds where lamb is an established part of the cuisine. Most of the lamb eaten in this country is consumed on the East and West coasts; 30% of the U.S. consumption is in New York State. Very little lamb is consumed in the western states, where most of the lamb is raised, although California has the second largest consumption, at 17%(1). An article in *Natural History* describes this phenomenon.

No one can say for sure why Americans don't all love lamb with the fervor of the Greeks and the Jews. ... It is speculated that a whole generation of GIs were subjected to Australian mutton during World War II and afterward wouldn't touch so much as a lamb chop. Then there is the fact that beef cattle were convenient machines for gobbling up the post-war grain surplus and turning it into cheap easy-to-prepare meat, while most sheep continued to eat plain grass. Not only that, Americans who did indulge in lamb limited their consumption to the most expensive cuts, the chops and the legs. As a result, lamb has an expensive image ...(2)

When you raise your own lamb, you can either take it to a meat locker for butchering and processing or do it yourself at home. Remember that if you slaughter lamb yourself, it is illegal to use the meat except for yourself, your family, nonpaying guests, and employees. You may not sell any portion of the carcass. An excellent and complete booklet by the Agricultural Research Service on home slaughtering of lamb, *Lamb: Slaughtering, Cutting, Preserving, and Cooking on the Farm,* can be obtained by writing to the Superintendent of Documents in Washington, D.C., or to your local extension service office.

The retail cuts of lamb are shown in Figure 10.1. As you learn to prepare lamb and learn your family's preferences, you will adapt your custom or home butchering to meet your cooking needs. How should you introduce lamb to your kitchen, your family, and friends? The worst way is probably to fry expensive chops quickly in a pan and set them on a plate. Lamb has an unusual flavor, and the newcomer should be carefully introduced to it, so try to make the unfamiliar familiar and not call attention to the fact that you are serving something new. The special flavor of lamb will come through most appealingly in dishes that are already favorites. Or, to quote *Gourmet,* "Anything that can be created with plain cooked roast beef is twice as interesting when done with lamb."(3)

Because our national consumption of lamb is so meager, we have not developed many traditional American lamb dishes. Instead, we have relied on recipes from cultures where lamb is an important part of the diet. So if your family members lean toward spaghetti and meatballs, you can use ground lamb in your favorite spaghetti sauce or in place of groundbeef in lasagna. If they like Mexican food, tacos and chili are also fine when made with ground lamb or stewing pieces. If they are strictly meat-and-potatoes types, some simple recipes like the ones included for Irish pasties or barley soup may suit them.

Developing a meat product that is palatable to the consumer is obviously desirable. But it is not easy to know what "palatable" is or how to achieve it. Some research has shown that both breed and type of feed can affect the flavor of lamb.

There is some research evidence that the finer-wool breeds may have a stronger flavor and that legume feeds tend to intensify flavor. But given an animal of the right weight and finish, most of the palatability is going

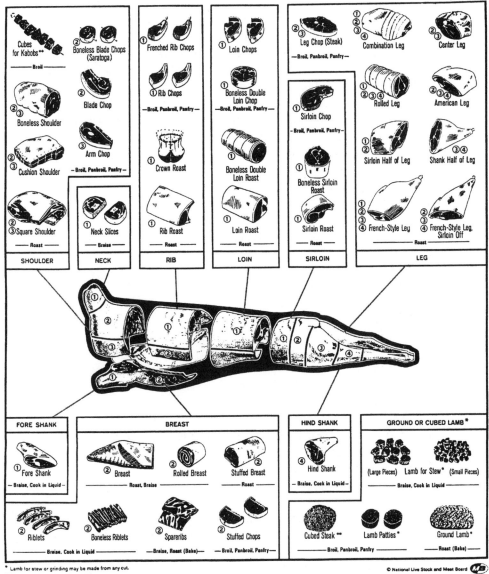

**10.1.** Retail cuts of lamb. (*Courtesy, National Live Stock and Meat Board*)

to be determined by the preparation. Cooking lamb at slow temperatures, not overcooking, and using some familiar ingredients can provide the best introduction. Following is a sampling of lamb recipes collected from personal acquaintances and from cookbooks. These recipes will be appealing to beginning lamb eaters and are ones that have been used with at least a few people who have never eaten lamb before.

The British Isles provides us with many ideas for cooking lamb, especially for using leftovers and less-expensive cuts in ways that will not surprise the meat-and-potatoes eater. Sherry Hanley, who was raised in Montana, uses a recipe for a pastry "sandwich" that was sent down to miners in their dinner buckets. It has been eaten in great quantities at the Hanley's St. Patrick's Day parties by people who swear they hate lamb.

## IRISH PASTIES

CRUST
1-⅓ C lard or shortening
4 C flour
Salt to taste
⅔ C cold water, approximately

Blend lard into flour. Add salt and enough cold water so that pastry forms a ball. Chill in refrigerator for at least 1 hour.

FILLING
1 to 1½ lb raw "lamburger" or finely chopped lamb meat
An equal amount of grated raw potato
1½ tsp salt
1 tsp ground pepper

Mix ingredients for filling. Roll the pastry dough to ¼ inch thickness. Cut 3-inch circles out of the dough. Place about a tablespoon of filling in the center of a pastry circle. Wet the edges of the circle with water and fold the circle in half over the filling. Seal the pastry edges with a fork. Place the pastries on a baking sheet and bake at 350° for 30 minutes, or until the pastries are golden brown. You can take them out before they brown completely and freeze them for later reheating. Serve them plain or with sweet-and-sour sauce or sour cream.

Not all recipes from the British Isles are quite so familiar. One woman newly arrived from Scotland gave special instructions for her custom lamb to be butchered because she wanted to make a traditional Scottish *haggis*—a sheep's stomach stuffed with pudding made of a mixture of various sheep innards and oatmeal. The trachea of the slaughtered lamb had to be still connected to the stomach to let steam out as the mixture simmered. As the instructions were written out for the butcher, he shook his head as if to say, "It's bad enough to eat lamb, but this is too much!"

Paul and Janet Abbas are both from Massachusetts families of Middle-Eastern heritage. The Abbases' buffet breakfasts are well known to their friends in Iowa, and lamb is usually featured. For one of the most popular dishes at these buffets, they brown ground lamb, onions, and pine nuts and season the mixture with allspice. They then add this mixture to scrambled eggs. It is an unusual combination, but the flavor is delicate and it looks familiar. Here is one of Janet's regular favorites.

## ROAST LEG OF LAMB WITH VEGETABLES

1 leg of lamb
1 lb of carrots, peeled and cut into strips
2 lb fresh or frozen green beans
4 onions, peeled and quartered
4 large potatoes, peeled and cut into 1-inch pieces
Allspice, about 2 tsp
Salt, about 2 tsp

Rub entire leg of lamb with allspice and salt. Place leg of lamb on rack in roaster and cover. Bake at 325° for 1 hour. Add water to roaster to cover the rack 1 inch and add all vegetables. Cover and cook approximately 1 hour more, until vegetables are tender.

The following recipe for Lamb Calzones, from the American Lamb Council, is a good example of using lamb in familiar Italian-American cuisine.

## LAMB CALZONES

**2 tsp vegetable oil**
**½ lb lean ground American lamb**
**2 cloves garlic, minced**
**½ C prepared pizza sauce**
**¼ C grated Parmesan cheese**
**2 tbsp diced green bell pepper**
**2 tbsp diced red onion**
**¼ tsp chopped ripe olives, optional**
**1 pkg (10 oz) refrigerated pizza dough**

Heat 1 teaspoon oil and saute lamb and garlic until lamb is no longer pink. Drain well and cool to room temperature. In a small bowl combine lamb mixture with 3 tablespoons pizza sauce, Parmesan cheese, pepper, onion, and red pepper flakes. Set aside.

On lightly floured board, roll out pizza dough to 14" × 10½" rectangle. Cut in half crosswise and into thirds lengthwise to make six 7" × 3½" rectangles. Brush half of each rectangle with remaining pizza sauce, leaving ¼" edges. Top each with ¼ cup lamb mixture and any remaining pizza sauce. Brush edges of crust with water and fold dough to enclose meat mixture. Use a fork to crimp and seal edges. Brush tops of calzones with remaining oil.

Place on baking sheet and bake in preheated 425° oven for 12–15 minutes or until golden brown.

Groups trying to promote lamb need to have recipes that produce mass quantities of food. Lamburger is the usual choice. The first two of the recipes that follow, all of which are from the *National Lamb Cookbook* (Shawnee Mission, Kansas: National Wool Growers Promotion Committee, 1980), may be handy for promotional activities.

## BARBECUED WHOLE LAMB

Choose a large lamb that will dress out to 65–70 pounds. Stuff the cavity with several loaves of day-old bread and lace shut, or cover well with foil. The bread collects excessive fat and helps to keep the lamb moist; discard it after the lamb is roasted. Allow 6–7 hours to roast, depending on the size of the lamb. Baste with the following sauce during the last half of roasting period:

> 1 large can tomato juice (46 oz)
> 1 16-oz bottle Real Lemon juice
> ½ C vinegar
> 1 bottle Worcestershire sauce (10 oz)
> 6 tbsp dry mustard
> 2 C brown sugar, firmly packed
> ½ C white sugar
> 10–12 cloves garlic, crushed
> 3 tsp freshly ground pepper

Mix ingredients, and put in a gallon plastic container or jug with a fairly wide opening. Use a new paint brush to dip into sauce and baste as directed. Make sauce a day before to allow flavors to blend. This recipe will amply serve 70–75 people.

## HOKEY POKEY
(Johnson County [Iowa] Sheep Producers)

> 1 lb ground lamb
> 1 lb ground pork
> ½ lb Velveeta cheese
> 2 tbsp Worcestershire sauce
> 1 tsp garlic salt

Brown meat, drain. Add cheese, Worcestershire sauce, and salt. Stir until cheese is melted. Serve hot on crackers. (Triple this recipe for a large crock pot.)

## LAMB STEW
(Alice Coicoechea, Elko, Nevada)

Have butcher cut up a front shoulder of lamb into small pieces or use lamb riblets. In a large skillet with a lid, saute 4 or 5 cloves of fresh garlic in a small amount of cooking oil in a heavy skillet. Salt lamb to taste and brown slowly, either on top of the stove or in the oven, stirring often. When lamb is browned, drain off excess fat. Add the following ingredients:

**¼ cup white wine**
**1 medium-sized onion**
**2 stalks celery**
**1 green pepper**
**2 or 3 carrots, cut into chunks**
**Pinch of rosemary**

Almost cover with water, put lid on skillet, and bake in slow oven, 325°, for at least 1 hour. Then if you decide you would like potatoes, add 1 or 2 cut into chunks. Let cook 1 more hour. The broth should be golden brown, which makes a very nice moist stew that can be served as a one-dish meal. If you do not care for the potatoes, add less water and serve over white boiled rice with a fresh green salad and garlic sourdough bread.

## FAVORITE LAMB AND BARLEY SOUP
(Makes 2 to 8 servings)
(South Dakota Sheep Growers Auxiliary)

| | |
|---|---|
| **1 tsp shortening** | **1½ C diced potatoes** |
| **1½ lb lamb shoulder, cubed** | **1½ C whole kernel corn** |
| **1 C sliced onions** | **1 tsp salt** |
| **1 C sliced celery** | **⅛ tsp pepper** |
| **1 medium-size green pepper,** | **¼ tsp thyme** |
| **diced** | **½ C barley** |
| **2 qt stock or bouillon** | |

Melt shortening in kettle. Add lamb, onions, celery, and green pepper. Cook over low heat until lamb is browned on all sides, stirring occasionally. Add stock or bouillon. Heat to boiling point. Add remaining ingredients. Cover and cook over low heat, stirring occasionally, for 45 minutes or until vegetables are tender.

## DAVIS COUNTY PARTY MEATBALLS
(Davis County, Iowa)

| | |
|---|---|
| 2 lb ground lamb | ½ C bread crumbs |
| 1 egg | 1 qt tomato juice |
| 1 pkg onion soup mix | 1 C sugar |
| 1 tbsp dried parsley | 28-oz jar barbecue sauce |
| Salt and pepper | 1 can beer |
| Chili powder | |

Blend first seven ingredients and form into small meatballs. Place in casserole. Add 1 quart tomato juice and 1 cup sugar. Bake 45 minutes at 350°. (This recipe can be frozen at this step.) Add a 28-oz jar barbecue sauce and 1 can of beer. Simmer until serving time.

## MEXICAN LAMB
(Mrs. Jerry Devlin, Terry, Montana)

1½ lb lamb breast or stew meat
Salt, pepper, garlic salt
1 small onion, chopped fine
4–6 C pinto beans, cooked or canned
1 can (10½ oz) condensed tomato soup
½ tsp chili powder
Oregano, optional

Brown meat in skillet. Turn frequently to melt excess fat. Remove lamb to a pan and sprinkle with salt, pepper, and garlic salt. Pour off all but about a tablespoon of the fat remaining in the skillet. Add onion to skillet and brown slightly. Add beans, tomato soup, and chili powder. Mix well. Place lamb pieces over surface of the sauce. Cover and cook slowly until meat is fork tender, about 45 minutes. Or bake in 350° oven for 1 hour. (Note: For a spicier taste, add more chili powder and some oregano.)

## LEG OF LAMB WITH MUSTARD COATING
(Rita Grafft, Olin, Iowa)

6 lb leg of lamb
½ C Dijon-type prepared mustard
2 tbsp soy sauce
1 clove garlic, mashed
1 tsp ground rosemary or thyme
¼ tsp powdered ginger
2 tbsp olive oil

Blend mustard, soy sauce, garlic, herbs, and ginger together in a bowl. Add olive oil by droplets, beating constantly, to make a mayonnaise-like cream. Paint the lamb with this mixture and set it on the rack of the roasting pan. The meat will pick up more flavor if it is coated several hours before roasting. Roast in a 350° oven 1 to 1¼ hours for medium rare.

## GYROS
(Eleanor Jones, New Mexico)

2½ lb ground lamb
3 tbsp chopped fresh mint
2½ tsp salt
Freshly ground pepper
2½ tsp oregano, crumbled

2½ tsp cumin
Tomatoes
Onion
Parsley, chopped

In a large bowl, combine ground lamb with the mint, salt, pepper, oregano, and cumin. Mix well until blended. Divide into eight even portions. Shape into logs about 6 inches long. Place under broiler for 8 minutes on each side (or until done as you like), turning once. Remove and place on center of pita round. Cover meat with tomato and onion slices and parsley. Spoon yogurt sauce over all. Wrap pita around burger and enjoy. (Sauce and pita recipes follow.)

## YOGURT SAUCE

1 small unpeeled cucumber　　　　　1 tsp olive oil
½ C plain yogurt　　　　　　　　　　¼ tsp salt
1 clove garlic, finely minced

Cut cucumber in half lengthwise, scoop out seeds, and discard. Finely chop remaining cucumber. In small bowl, mix cucumber with yogurt. Add garlic, olive oil, and salt. Cover and let stand in refrigerator for 30 minutes to blend flavors. Makes about ¾ cup.

## PITA ROUNDS

1 pkg hot-roll mix　　　　　　　　　1 egg
¾ C warm water　　　　　　　　　　2 tbsp vegetable oil

Pour warm water in a medium-sized bowl. Sprinkle yeast from roll mix over this. Stir until dissolved. Beat in egg and vegetable oil until blended. Stir in flour mix from package to form a soft dough. Cover bowl with clean towel and keep in warm place away from drafts about 45 minutes, or until dough has risen to about twice original size. (At this point turn on oven to 450°.) Punch down dough in bowl, kneading lightly with hands until it is stiff and elastic. Divide into eight even pieces. Form eight smooth balls by rolling dough between palms of hands. Place two balls at opposite ends of a lightly greased baking sheet and flatten evenly into 7-inch circles. Bake on lowest rack of preheated oven for 6 minutes or until lightly browned. Remove breads to wire rack to cool. Repeat baking until all are cooked, cooled, and stored in plastic bags. You may bake pitas ahead. Before serving, take pitas from bag and sprinkle each lightly with tap water. Wrap in foil and heat in 350° oven for 5 minutes until warm and foldable. Makes 8.

The following recipe was a recent winner in a lamb barbecue cooking contest sponsored by the Lamb Council of the American Sheep Industry Association.

## JALAPENO MARINATED LAMBCHOPS

**8 well-trimmed lamb loin or rib chops, cut 1 inch thick**
**¼ jalapeno pepper, seeded**
**1 large clove garlic, halved**
**1 small onion, quartered**
**1 piece fresh ginger, peeled (about 1 inch)**
**2 tbsp each honey, orange juice, peanut oil, and soy sauce**
**⅛ tsp coarsely ground black pepper**
**Orange slices and cilantro, optional**

Combine jalapeno pepper, garlic, onion, and ginger in blender container. Process until finely chopped. Add honey, orange juice, oil, soy sauce, and black pepper. Process until blended. Place lamb in plastic bag, add marinade, and turn meat to coat. Close bag securely and marinate in refrigerator 15–30 minutes, turning occasionally. Place chops on grill over medium coals (hold hand about 4 inches above coals—you should be able to count to 4 comfortably). Grill for 9–10 minutes for medium (internal temperature 160° on meat thermometer), turning once. Garnish as desired. Serves 4.

These examples provide some ways to fit lamb into familiar cooking routines. The industry is producing an increasingly high-quality and healthy product. You can help publicize this through your own efforts in the kitchen.

## REFERENCES

1. American Sheep Industry Association. *Sheep Production Handbook*. Englewood, CO: ASIA, 1992, p. 404.
2. Raymond Sokolov, "Fleece Afoot," *Natural History* (October 1980), p. 94.
3. "Lamb Legacies," *Gourmet* (April 1980), p. 42.

# 11. On Location

You will be raising sheep in a specific area of the country and probably with a specific focus in mind. This book emphasizes the value of contacting and getting to know experienced shepherds who have been working under similar circumstances. But it is also useful to understand something about the history and scope of the sheep industry and how and why sheep are raised in other parts of the country. The following stories offer a bit of this journey.

The first two stories, "An Island Ideal for Sheep," and "If You Take My Sheep, I Will Die," present two different settings for the long and important history of sheep in this country—one on the East Coast and one in the Southwest. At roughly the same time, the domestic sheep became a way of life in both places, and today's shepherds are still tied to those early roots.

While the beginner is not apt to be starting with a range flock, Colorado families Spence and Connie Rule in Brighton, and the Browns in Ignacio, offer a picture of western sheep production. Still using traditional herders and mountain grazing lands, they bring intense management and concern with consumer needs to their traditions.

Bill and Jan Butler of Whitehall, Wisconsin, and Randy Irwin of Freeland, Maryland, give a picture of some new approaches to niche marketing of sheep products. Bill and Jan are part of a growing number of sheep producers who focus on dairy products, while Randy has developed some interesting, and sometimes exotic, outlets for the wool that is the focus of his operation.

Tom and Sandy Clayman of Hutchinson, Kansas, and Mary Ann Nipp of Arlington, Nebraska, are breeders of purebreds with an interest in developing sheep that can contribute to commercial productivity. While the showring and "pretty sheep" are of great interest to them, they also share with the beginner an interest in genetics as the key to a successful operation.

Stan and Jean Potratz, of Washington, Iowa, are in the sheep equipment business and also have a flock of 300. They share their ideas about that most important piece of equipment in your new operation—the sheep themselves.

## A Tradition on Martha's Vineyard: An Island Ideal for Sheep

The island of Martha's Vineyard—90 square miles of grass meadows, hills, and woods fragrant with scrub oak, pine, and salt air— lies 3 miles off the coast of Massachusetts. The land meets the ocean in varied degrees of drama, from the high, wildly colored cliffs of Gay Head to the plains outside the town of Edgartown. Close to East Coast metropolitan areas, the Vineyard's population of 12,000 swells in the summer to some 70,000, a summer residence to entertainment, literary, and political celebrities.

But there was a time, still within the memory of older residents, when the Vineyard was a self-contained farming community with sheep as a critical part of the economy. Today, there are only about 200 sheep on the Vineyard. But a century ago there were thousands, with a genetic line going back well before the American Revolution. Sheep were a major part of this island's economy, politics, and social life.

Before the Pilgrims landed at Plymouth Rock in 1602, the Wampanoag Indians on the island were joined by an English sea captain, who named it Martha's or Martin's Vineyard. In 1642 Thomas Mayhew became the first British governor. Sheep were an important part of this first settlement, and Governor Mayhew hunted for a master weaver who could help turn the Vineyard wool into cloth.

With the English establishment on the Vineyard, conflict and confusion soon developed between the English concepts of land ownership and the long-held communal ownership of the Wampanoag tribal lands. The English government "assigned" to the Indians certain areas on the island that were diminished through various ways in the quest for more grazing land. One means of access was leasing arrangements, expressed in terms of "sheep rites." For example, an 800-acre farm at Gay Head was leased to Peter Norton in the late 1700s for mutton, wool, and leather, which was used in his shoemaking business.(1)

The census in 1766 showed the flourishing Vineyard with close to 3,000 people, well-stocked herds of cattle and sheep, a vital cod fishery, and 15 or 16 whaling ships. At the time of the American Revolution, the two principal industries on the Vineyard where whaling and sheep raising. In September of 1778, a British general, Sir Charles Grey, anchored 12 warships and 20 transport vessels off one of the island towns. He "requested" food for his troops who were stationed in Newport, Rhode Island, and promised to pay. The Vineyarders had to bring 10,576 sheep down to the docks.

> With the war began a series of calamities....The vessels were all taken and destroyed—their young men were captivated and many of them died on board prison ships. General Gray in the month of September 1779 carried off at one time a hundred and twenty oxen and ten thousand sheep, leaving on the island only about four thousand of the latter ... the sheep were appraised at two dollars. But two years later no more than one third of the sum which was due had been collected. From that depressed state, the island has not yet recovered.(2)

Interestingly, it appears that the sheep belonging to the Wampanoag tribe members were not touched during the raid.

The Vineyard flocks were far from wiped out, however. In 1809 David Look established a woolen mill, with the main product a heavy, tightly woven fabric called satinet. Island farmers brought their wool to Look's factory. The satinet was advertised in the *Vineyard Gazette,* "Those who have taken my goods for a long voyage have made the discovery that one half the number of garments made from my clothes will answer the purposes of a voyage." Apparently the wool satinet provided warm clothing for the sailors on the whaling vessels. At its peak, the factory employed some 60 people, all working with island-grown wool. The business seems to have died out with the death of the whaling industry and the advent of petroleum. The old factory building is now owned by the Martha's Vineyard Garden Club, and the wide pine floors show smooth indentations where the women workers stood at their looms.

Several of those few people raising sheep on the Vineyard today have traditions that go back to those early days, and their sheep are penned by the same picturesque, lichen covered stone walls. There is a determination to maintain that tradition. Allen Whiting, his sister Prudy, and his brother Danny are part of a family of shepherds that goes back to those

**11.1.** Allen Whiting continues the shepherding tradition.

first sheep to arrive on the Vineyard in the early 1600s, making them the tenth generation of shepherds on the island, with their children being the eleventh. Their father, Everett Whiting, who died in 1980, recalled his boyhood when much of the island was sheep pasture. What is now woods was all clear. Four-sided stone pens were used to keep the sheep that strayed from home until their owners would claim them—the sheep were identified by an old English system of ear notches. Everett himself once had a flock of 300. Today, with a stubborn insistence on the tradition, even in the face of the powerful tourist economy, Allen maintains a family flock of about 35 Cheviots. Prudy runs a bed-and-breakfast in a house on the farm where she and her brothers grew up. The house was built in 1668 by Josiah Standish, son of Miles Standish. The living room of the old house looks over a beautiful slope down to a pond, and the back entrance looks over the field and the barn where the sheep often graze.

Allen, who is an artist, lives and has a gallery in the Parsonage, a large Victorian house next to the bed-and-breakfast. His wife, Lynn, and children share in the work of both the farm and the gallery. Prudy and Danny and their children are all participants as well. The farm figures prominently in Allen's gentle but powerful painting, and sheep are not infrequent subjects.

"It's ironic," Allen says. "Once it was not uncommon for a farmer to help out a starving artist by letting him use a loft with good light in a barn in exchange for some hard labor. Today, I guess it's the art that's supporting the farm."

Hollis B. Engley of the Dukes Cooperative Extension Office on the Island has the records of the early agriculture days of the island. "The sheep numbers have been steadily going downhill since about 1779," he says. "And the countryside has become heavily overgrown with either scrub or tree growth. Being isolated from predators, the island was ideal for sheep, but there is far less interest in sheep now."

But here and there, the island sheep are still taking advantage of this perfect home under the care of shepherds who believe not only in the value of their products, but in the tradition they represent.

## If You Take My Sheep, I Will Die: A Navajo Way of Life

> When she was sixteen years old, they had the maiden ceremony for Changing Woman, called Making-the-Path-of-Life-Beautiful. They dressed her in white shell shoes, fine deer skin robes.... Her hair was parted in the middle and hung down tied at the back half way to the ends. They invited Man-Wrapped-in-a-Rainbow and he came and brought many shell dishes and food, and he also brought her a baby lamb and all the gods came.(3)

Far from the fog and salt air of New England, but at roughly that same time, in the sunlit spaces of the Southwest, another form of shepherding took shape in the New World. The history books say that the domestic sheep were brought in by the Spanish conquistadors in the early 16th century. In 1598 it is recorded that Juan Onate colonized the area of the Southwest where the Navajo lived—Utah, Arizona, New Mexico. Unlike many of the introduced elements of Spanish and Anglo culture, the sheep became a welcomed part of Navajo life. They used sheep for food, hides for warmth, and wool for weaving, which became a major art form and trade commodity. In the Navajo story, the sheep have been with the People since the creation.

> "Sheep were security for the People," says historian Lynn Bailey. "Mutton fed their families, wool kept them warm and provided cash from the traders. As shepherding was a cooperative enterprise reaching far beyond the nuclear family, it served as a symbol of social integration. Children were given livestock at an early age, introducing them into the herding activities of the group. No Navajo would think of marrying without bringing sheep to the herd of his wife's family."(4)

"It is in the sheep herd," comments anthropologist Cary Witherspoon, "more than anywhere else, that the divergent interests of the individual members of the unit are converted into this meaningful and cooperative endeavor. Sheep were the foundation upon which rested Navajo kinship solidarity."(5)

But the history of the Navajo and the sheep, like the history of the People as a whole, is fraught with conflict and pain, periods of destruction and repression, followed by efforts to rebuild, all by the same government. The sheep that acclimated so well to the terrain and to the Navajo people were the Spanish Churro, ideally suited to the arid conditions of the Southwest. One special feature of these sheep was the wool, a long, coarse outer coat and a short, fine inner coat that is highly valued by the weavers. But the original Churro animal, so well adapted to the environment, was weakened by the introduction of "improved" breeding stock with wool that did not meet the weavers' needs and that was unable to adapt to the local conditions. The Churro also went through some systematic extermination.

The Navajo were forced from their lands in 1863 in the Long Walk, led by government troops, and were in captivity in Bosque Redondo for 4 years. Many of the Navajo died of starvation and disease during that period, and the sheep population was also devastated. When the Navajo were finally released from captivity in 1868, they were given back this center of their economic and social life. With the remains of their flocks and a "gift" from the government of two sheep for each person, the Navajo went back to a pastoral lifestyle, following the sheep to various grazing lands. Their reinstatement was so successful, that by 1930 there were a million sheep, causing a new set of government interventions because of overgrazing.

To save the depleted rangelands, the government stepped in to reduce the flocks. The Navajo were not compensated for their loss, and often the agents did not speak the Navajo language. The event was devastating for many of the families, economically and socially.

"If you take my sheep," one Navajo herdsmen said, when confronted by an agent, "you will kill me. So kill me now. Let's fight right here and decide this thing."(4)

With both the Churro breed and the shepherds themselves affected by the reduction program, it is a miracle that there is any sheep industry left at all in the Navajo culture. But there is definitely concern about its long-term survival.

Bill Malone, who runs the Trading Post in Ganado and whose wife is Navajo and a weaver, puts it this way. "The real thing is the entire cycle for the Navajo. It was part of the year ... lambing in the spring, raising the sheep on the range in the summer, culling in the fall. They ate lamb and mutton, they used the pelts to sleep on, they sold some lambs to pay off a grocery bill. What I see in the future is that if it isn't a special interest, a love for sheep, it's going to be hard to stay in it. It isn't cheap to raise sheep. And you can't let them roam free on the grazing lands.... You need to have a herder. And they're hard to find. No one wants to be a sheep herder now. Many of the women have outside jobs, and they were the ones who traditionally stayed with the sheep. I've noticed a transition to cattle even though it's not traditional, because the cows won't be at- tacked by coyotes. The people that are really into sheep now are the strong weaving families. Navajo wool always got a low price because they would save the best wool back for weaving."

Interest in Navajo weaving as art dates back to the 1960s. The Navajo themselves do not put the rugs on their floors. They are used for trade. The rugs are sold to get the bills paid. "My wife weaves, she is weaving a rug now, a Storm Pattern," says Malone. "She's going to try to make sure each of our five kids gets a rug. Lots of families who weave don't have any rugs saved."

Twenty years ago, Dr. Lyle McNeal, then at California Polytechnic State University at San Luis Abisbo, began a quiet obsession to ensure the survival of the Churro breed and the sheep industry on the Navajo reser- vation. It began with drives through remote areas of the reservation, look- ing for sheep with a Churro look. These efforts have resulted in a sub- stantial breeding stock that is being given back to Navajo families. The Navajo Sheep Project has also involved a variety of activities to promote better practices by the shepherds and to develop markets for the prod- ucts. Currently, a grant from the U.S. Department of Agriculture, as part of the Sustainable Agriculture and Research and Education (SARE) pro- gram, is helping provide technical assistance and other support to a small number of families.

Joe Benally and his artist wife Carol are part of the project. They live in a hogan they built themselves, full of sunlight and tile and the artifacts of their involvement with sheep—piles of wool and yarn in the varied earth tones that the Churro produces. The flock is grazing over at Joe's parents' land, but here the Churro rams are penned near the house, as are some horses and a fierce-looking black llama, who is supposed to guard the flock but seems to have earned the nickname "Dances on Ewes."

"One of the first things I remember," Joe says, "is when my mom gave me a sheep. She said 'Here, it's yours. Take care of it.' It's almost like a test for the kids to see if they take to it. I guess I did. We grew up on goat milk and sheep milk. The sheep have always been with us. The spirit of the sheep has always been here. It's kind of weird that we have to rebuild what was always here. We have about 50 sheep now. We sell lambs and we sell sheep for meat. People like the sheep here, the Churro tastes good. We have people who could buy blackface sheep to eat for 65 cents a pound, but they come here and pay 90 cents because it tastes better."

Joe and Carol, who are both teachers, are, like all sheep producers today, working to identify markets for their meat and their wool. They have developed a catalog of products from the Churro, along with spinning and weaving equipment under the title "Iina Ranch," Iina meaning "preserving the Navajo-Churro sheep as a way of life."

The Benallys' concern about preserving the sheep industry is shared by many in the Navajo Nation. In September 1995, the First Navajo Nation Sheep Producers Summit was held in Farmington, New Mexico, with the theme "Saving Our Sheep Industry." The concerns facing producers include the issues of finding better ways to market both the wool and the lambs and dealing with grazing rights. "The issue of the grazing rights is very complex," says Al Grieve, of the Trading Post in Pinon. "My wife's family used to run 600 head of sheep in the late '60s and '70s. They were strong livestock people. The sheep permits are passed on through the women. It used to be that a woman would pass on the permit to a daughter, but now there are so many people involved, and there is not enough to go around. And, there is still a problem with overgrazing."

Sarah Natani and her husband Leo have a ranch within sight of the mysterious, looming Shiprock. Their sheep are a mixture of the Churro colors of browns and reds and grays and white that match the colors of the open range where the flocks still roam. Sarah is a weaver herself, and their hogan, close to the highway, is filled with weaving supplies and her own rugs. Her children are grown; she is concerned about the preservation of the traditions, that there may be no young people interested in taking care of the sheep.

Irvin and Marjory Curley seem to be addressing this concern. They live with their children on a ranch near Ganado, Arizona, still and remote, with the sounds of the wind in the pinons mixed with the bleating of the sheep. The flock, now about 150, is all of the white variety of Churro. The Curleys have been working to improve the breeding, making use of some of the stock from the Navajo Sheep Project for wool production. The sheep are a family affair. Although Marjorie and Irvin both have jobs in town, Marjorie with the school system and Irvin with the phone company, the family year is still influenced by the sheep. In the summer they move to the grazing lands on the nearby Defiance Plateau, making a camp to live in.

**11.2.** Sherilyn Curley tends the Churro flock.

Fourteen-year-old Sherilyn Curley is playing an active role in the shepherding. "One of my main jobs is feeding in the winter, and helping with the lambing. I go out and get the lambs and ewes penned up together. This last spring we had a big snow and the lot was very muddy, and there were a lot of problems. I took a week off of school to get it cleaned up."

The wool, highly valued by the Navajo weavers, is receiving special attention in the Curleys' flock.

"Every year we measure the wool," Sherilyn explains. "We measure from the pink skin to the end of the wool, without stretching out the fiber too much. We also look to see about wool on the face because it's not too good to have a wooly face. Last year I just caught the sheep, but this year I got to do the actual measuring." Sherilyn doesn't have a future career in mind, but she would like to keep the sheep when she grows up and continue the tradition.

"I'm trying to learn to weave," she added. "I haven't gotten very far yet, but I'm going to learn from my mother."

## Searching for Plate Coverage

The sheep operation of Spence and Connie Rule, Rule Feedlot, Brighton, Colorado, covers quite a space. It goes from grazing land in Utah, where their flock spends the winter and then lambs in the spring, to the summer grazing lands in the Rocky Mountains, to their feedlot in Brighton, Colorado, to their packing plant, Denver Lamb, 45 minutes from Brighton. But the narrow wooden loading chute coming out of the feedlot corrals is, according to Spence Rule, the most important space in the operation. Here Spence and maybe two or three other workers sort and make judgments about the carcasses of over 500,000 lambs a year. These lambs will leave Denver Lamb as packaged products for the grocery store meat counter. From the genetics of their own ewes on, the Rules are focused on making that product attractive to the consumer.

"We started out as lamb feeders," Spence says. "We either fed custom lambs or we fed our own. The feedlot was our profit center. But the times were changing and the packing business was changing. In 1980, we had a lot full of lambs ready to sell—35,000 head—when both packinghouses in Denver, Monfort and United, went on strike. We had no place to sell the lambs. There was an old cow kill in Denver that was empty, so that's how we got started in the packing business." As Denver Lamb grew, the profit center shifted. "Now we see the feedlot as a quality-control center for the packing plant. The main thing for us is to have the same product day in and day out. Some packers will sort carcasses in the cooler, but we do it right here in the feedlot."

The feedlot is a simple operation with large, wooden corrals with feed bunks, some covered space, space for feed storage, and a place to shear. And lambs everywhere, arriving in the lot year-round to keep the supply steady, with usually about 45,000 in residence at any one time. Spence and Connie manage the operation with five additional employees.

The ideal market lamb is different now than it was when the Rules started in the packing business. Then, a lamb weighing 95–110 pounds

**11.3.** Spence and Connie Rule—still shepherds first.

was the goal. Now the Rules like lambs that weigh 125–140 pounds. But, with the leanness that consumers demand today and being the ones who have to trim fat off in the packing plant, they watch the lambs closely to make sure the fat doesn't go on in the first place. "We would be paying a feed bill to put fat on, and then paying someone in the plant to trim it off," Spence smiles.

"We want bigger loins and bigger legs," he adds. "The restaurant trade calls it 'plate coverage.' The chops need to be look more like a steak. With a 95-pound lamb, you have to put three or four on the plate."

Over the last 15 years, there has been a dramatic decrease in red meat consumption in the United States. Although lamb consumption has not gone down percentage wise to the same degree as other red meats, lamb already had its long-term consumer prejudices to deal with. Spence Rule is finishing a term as chair of the Lamb Council of the American Sheep Industry Association. The Council spends both time and money on advertising to overcome these prejudices on behalf of sheep producers.

"Our whole focus has been 'Do it quick, do it simple,'" Rule says. "That's true for all the meat promotions now. But because lamb is not as familiar to American consumers, we have to add the message, It's different! Take a risk!"

Last year the lamb council did a promotion through *Better Homes and Gardens* that reached 20 million readers. There were recipes and coupons to redeem at local supermarkets toward the purchase of lamb products. There was, according to Spence, a good response. In searching for promotional recipes, the council has focused on using familiar products, such as soy sauce, mustard, and barbecue sauce, as opposed to exotic cuisines.

At Denver Lamb, the Rules try to put the familiar, fast, and easy message into developing and marketing their own products. Over 60% of their kill is packaged at the plant, rather than going to "breakers"—the business that cuts up the carcasses for the supermarkets. With people wanting things for quick dinners, the traditional 6-pound leg of lamb has less appeal. Denver Lamb takes steaks off the top of the leg and then markets a small roast, 2½ pounds, that is premarinated and ready for the oven. Preseasoned fajita strips and marinated loins are also popular. Over 50% of their meat goes to the eastern distribution points of Boston and Philadelphia.

For the Rules, the message for beginning sheep producers is the same one they have for themselves. "Genetics is key," Spence says. "We need a lamb that will gain. There is a market for every kind of lamb, but the heavier ones are easier to market. If you are working with a lot of Finn breeding, you have to pay attention to your marketing and get the lamb up to maybe 120 pounds. This will take longer." Connie is adding to the genetic pool with one more sheep project, her own purebred Suffolk operation. "It's a hobby," she says with a grin.

The Rules are very interested in consistency in their feedlots. Buying lambs that are of roughly the same type is helpful. This is easier to do in the West than in the Midwest, where the flocks are smaller and more diverse. "And," Spence adds, "there are more producers and more buyers in the Midwest, all with a different idea of what a fat lamb should be."

The Rules' own range-flock operation of 5,000 ewes leads back to some old traditions, with herders managing bands of 1,200 ewes each. The herders now are primarily from Chile, Peru, and Mexico and are hired through a company that takes care of getting the work visas and other legalities. The camps are on wheels and the herders often have their families with them. They stay with the sheep all year, with a month to 6 weeks off, moving from Utah to the summer range in Colorado and back again. They are geared for a 105% lamb crop, because of the losses to predators and exposure. The ewes with twins may be hobbled or tied to a bush to allow the lambs to nurse, but there are no jugs or pens.

From lambing time in April to September, the lambs stay with their mothers with little interference, other than docking and castrating, which takes about a half a day for each band. "This year we used elastrators for castration," Spence says. "I don't know why we changed really. We've always used the traditional castration with the teeth. Most of the range producers still do that. It's clean. It keeps infection down. It has worked pretty well."

In mid-May, the flock is moved to summer grazing in Colorado. This past summer was an especially good season, with good rain. The grasses head up late in the summer, providing a form of grain that the lambs can use well. Many were ready for the packing plant when they came in off the range in September, with no additional grain feeding. "They can do real well up there," Spence says. "If they weigh 110–120 coming off the mountain, I don't keep them in the feedlot. They get enough fat on them in a year like this, and you can't tell any difference in the carcass between them and the ones that are fed in the lot."

The Rules' health program is a basic one. Ewes are vaccinated for vibrio and rams are tested for semen and epididymitis. The lambs are wormed when they come into the lot and treated for external parasites. Antibiotics are used to prevent shipping fever and are added to the ration midway in the feeding process.

As packers, the Rules are especially sensitive about injection sites for health treatments. Connie emphasizes that injections should always be done in the neck or in the "armpit." "If you vaccinate in the shoulder or leg, you go right into the muscle," she says. "This can form abscesses that have to be trimmed out of the meat. And with the pelt, an abscess can leave a hole. If an abscess forms in the neck, it's out of the pattern."

The Rules' approach to feeding has evolved into a formula that they don't like to tamper with. They expect a gain of 0.6 pounds a day and keep the lambs for a 40- to 45-day period. The ration is built around a staple of beet pulp from a local mill and hay pellets bought from Colorado and Wyoming. To this ration they gradually add straight combined corn, starting at 5%, working up to as high as 75%. "We find we're still getting more red meat than fat," Spence says. They watch the lambs' adjustment to the grain feeding carefully, mostly by the manure or by whether corn is being left in the feeders.

With the large number of lambs on feed at all times, Spence has found that shifting the ration when less-expensive alternatives come up—least cost rationing—can result in expensive adjustments for the lambs. "We always have a bunch of lambs in the middle of their feeding time. By the time you change the ration and get the lamb adjusted and back up to speed, you've lost time and money. I wouldn't change the ration now unless I thought I was going to keep it that way for a long time."

The Rules feed lambs for Coleman natural foods, as well. This requires a separate feeding setup and no use of antibiotics. The price paid for the lambs is higher than the regular feedlot, but it is not overall more profitable, because they find it costs more to raise them.

Wool, while a relatively minor product, is well tended in the Rule operation. Spence feels that the shift from the wool incentive program may actually end up being a good thing. "I think it contributed to bad management and a bad product," he says. "Because the payment was always there, no matter what the quality of the wool. Now we have to pay more attention to it."

The pelts on the unshorn lambs have been more profitable for those with better skins, but if the skins aren't good, then it's better to shear. The Rules shear about 250,000 lambs a year, bringing in a shearing crew from New Zealand. They have built a new shearing floor in a barn, with chutes coming up from the lots. They also have a wool baler that packs the wool into large burlap bags. This makes it easier to market, especially for possible shipment over seas.

Despite the "vertical integration" of the Rules' business and the large numbers of lambs that go in and out their gates every year, the operation is still about the basics—about animals being fed and cared for so they can produce a valuable product.

"I'm still a shepherd," Spence says. "That's what I like best."

## A Western Romance

For an American romance about sheep production, the setting could be in Ignacio, Colorado, outside Durango. The main characters would ride horses and trail thousands of sheep up to the wilderness summer grazing lands high in the mountains, and they would sleep under the stars in their bedrolls, and fight off bears that attack the sheep. Their ranch home would be in high, golden fields, spanned by the biggest of blue skies, with the mountains dark in the distance. And they would care for the sheep, and even with the hardships and sometimes unpredictable rewards, they would love their work.

Jean and Casey Brown and their son and daughter-in-law, J. Paul and Debbie, could play the lead roles in this romance—except that they would probably be too busy. They would actually be taking supplies by horseback up to the herders who care for their flocks in the mountains, or going to producer meetings, or providing leadership in the sheep industry business. They might be moving sheep or catching up on records. They would be involved in an intense period of lambing, and later they would be working on marketing and promoting their products.

Casey and Jean have been in the Ignacio area for 37 years and currently have a flock of 2,300 ewes. Their summer grazing is in the mountains, a 2-hour drive and a 25-mile horseback ride from their home. The sheep are trailed up to the grazing land on June 20, and between then and October when they come back, weekly trips have to be made to take food for the herders and salt for the sheep, and to help the herders move the camps.

"They use mules to carry the supplies and move the camp," Jean explains. "Casey uses five mules and one lead horse. We don't truck the sheep at all, except when the lambs go to market. They come back here and then go on to the winter range in New Mexico. It takes about 4 days from Ignacio to get to the winter range. The sheep are moving all the time, but they don't get stressed. We bring them back for lambing late in April."

The centerpiece of all the movement is the federal grazing land. The grazing program was started with the Taylor Grazing Act of 1934. The leasing rights were handed down in families, but when someone left the business, they sold their rights.

"When Casey's dad was in the business, there were 20,000 sheep up in the summer grazing land where we go now, " Jean says. " Now with J. Paul's and ours, its only 3,000—we're the only ones up there. I would have worried back then about overgrazing, but not now. Some people think the ranchers have taken advantage of the government. But I see the pro-

**11.4.** The Browns' sheep on the move from the summer grazing land (*top*); Debbie Brown on the Ignacio, Colorado, ranch (*bottom*).

gram as beneficial. The sheep eat weeds. The old forest dies out. Sheep help with the reseeding."

The sheep stay on the summer grazing land until October. Then in November they go down to the winter grazing lands in New Mexico. The bucks are put in with the ewes down in New Mexico on December 10. "It's a magic day," Jean says with a laugh. "We can't do it a day before or a day after. Every other day we add some bucks until they are all out."

The ewes are brought back to Ignacio from New Mexico in April for an early May lambing. While in some ways the Brown operations can be considered range flocks, their lambing is done on their home place, not on the range, and is more intensive than in a traditional range flock.

"We really work at the lambing," Debbie says of their 1,000 ewe operation. "We have the ewes in pens, about 45 ewes in each. And there are lambing jugs all along the sides. We put the ewe and her lambs in the jugs to make sure they're sucking and then we paint brand them. Each ewe is tattooed and ear-tagged, and we record whether she has a single or a twin. If she has a single twice, we get rid of her, except for the yearling year.

"The first year we moved here, we didn't have corrals, so we just lambed out," Debbie says. "It didn't work out as well. We save a lot more lambs now. We went from a 150% lamb crop to a 190%."

"We have our own breed," she explains. "Its about seven-eighths Rambouillet and one-eighth Finn. We call them Rambouilinns. We've closed our herd and we just use our own replacements for both ewe lambs and buck lambs, and the only buck lambs we use are triplets. We monitor closely for problems with inbreeding—if there is a problem, we put that ewe with a blackface buck the next year for market lambs. We do a lot of intensive recordkeeping. The year we started recordkeeping was the year we made the big jump in our crop. We took out all the ones that were consistently having singles and that made a difference. This year we had 186 sets of triplets."

Debbie and J. Paul plan for 20 days of really intensive lambing. "We are out there the whole time—we try to go in for sleep around midnight and then get back up at 4. First thing in the morning we have two of the boys help." The Browns do a lot of hands-on management to make sure every lamb has a mother to nurse with. "I'll see J. Paul down in the jug with his feet propped up with a lamb, and he's got that lamb there and he squirts a little milk into her and that's how they start sucking," Debbie says.

"My job is to take a lamb that's not being fed and find a ewe that looks like she's having a single, and while she's lambing, I'll get that orphan on the ewe. I get the lamb underneath the new lamb and mix them up and 99 times out of a 100 she'll take it. Over the years I've learned where the ewe smells the lamb the most—like under the tail and on the sides and over the back."

Debbie and J. Paul have four sons who are part of the work crew. The boys have their own flock of Suffolks for 4-H projects as well.

"Our main income comes from market lambs," says Debbie. "This year we contracted with a feedlot in Oklahoma before they were born. We had five or six people bid on our lambs this year. The lambs are about 85 pounds when they come off the mountains in October, which we think is pretty good for lambing as late as we do."

Health concerns are the normal ones. The sheep are wormed when they come out of the mountains, and then again before they go to the winter grazing lands.

"Up in the mountains it rains every day and we sometimes do have a problem with hoof rot," Jean says. "Not a big problem, but I actually make little booties from old Levis. I put socks on them to hold the powdered copper sulfate on. We don't have facilities for a foot bath up there."

While wool is a small percentage of the income for the Brown's, they have always tended it carefully and are being even more attentive now. This year Debbie and J. Paul used an Australian packer, a hydraulic packing process, instead of tying their wool. And they're looking into skirting the wool, where the fleece is laid out and the britches and the bellies are taken off, which is supposed to increase the value. Anything that is under 64 (the count system of measuring fineness) is taken out and marked separately.

"We actually look at every fleece," Debbie says, "And if its too coarse, we put that ewe in with a blackface buck for market lambs. We try to keep our wool really fine."

Because they work hard to produce good lamb and wool, marketing and promotion efforts are important to the Browns too.

"I don't understand what the problem is with the United States and lamb," Jean says. "It's the preferred meat of the world. And the discrepancy between what the consumer has to pay and what the producer gets paid is way out of line with other meat producers."

"We've been doing a lamb-tasting day in October for several years," Casey says. "One year we had a lamb-tasting in every City Market store in Colorado. Jean put me in the one in Cortez. I did it on a Saturday, and I remember one of the store managers said, 'You can't sell lamb in Cortez.' Well, I had little grills set up with bite size pieces of lamb and I'd put the bites on toothpicks. When the guy came by with his basket, I didn't say 'Would you like to try some lamb?' I'd just hand him the taste. He'd go by and then look up at the sign and say, 'That can't be lamb—that's delicious!' Another time we cooked preseasoned legs and offered samples. The first time they sold out of the legs by 1:30. The next time they ordered three times as much and sold out again by 1:30. The point is, it's a good product. This kind of marketing works."

Two years ago, Debbie called all over the country to branch out and try to find new places to go with the lambs. "I called Armour and a bunch of other packers to see if we could do better." Jean has checked out the wool festival in Estes Park, Colorado, too, to see what kind of wool the hand spinners and craft people were interested in.

Both J. Paul and Casey Brown have been active in industry activities. J. Paul is the president of the Colorado Wool Growers, and the whole family is active in the American Sheep Industry Association.

"Our way of raising and running sheep is sort of like the past," Jean Brown says. "And, I do think sheep-raising is romantic. We love it."

## Close Encounters with the Third Crop

North of La Crosse, Wisconsin, in the rolling hills of Trempealeau County, the lush green pastures and fields of recently picked corn are sprinkled with the sharp black and white pattern of Holstein cows. As the sun sets, and the hills grow dark behind the farmsteads, lights shine from the glass block windows of the milking parlors with warm familiarity. This is, after all, still America's dairyland.

North of Whitehall, on the picture-perfect valley farm of Jan and Bill Butler, America's dairyland takes on a slightly different look. The rolling pasture on this farm is not dotted with Holsteins, but with sheep, and the milking parlor has a ramp up to elevated stanchions where these small milk producers munch on their evening rations, while Jan and Bill Butler do their work.

In another country, the sight might not be so unusual. France produces over a million metric tons of sheep milk a year; Greece and Italy produce over 600,000 metric tons each; and Turkey produces 893,000 metric tons. The United States actually imports 16-million pounds of sheep-milk cheeses a year.(6)

"Well, yes, I guess we are an oddity here," says Bill Butler with a grin. "When we bought this place 3 years ago, it had been a dairy farm, a cow dairy, for about 120 years in the same family."

Jan Butler adds, "One of our neighbors was saying the other day, 'Remember when you first moved in, all the traffic that was going by?' People couldn't quite believe it. Milking *sheep?* We still have people just dropping by to see the operation, seeing what we're doing. They're curious."

The Butlers' sheep dairy experience started back in 1988 near Albany, New York. They were living near Jan's parents' 12-acre home and had a few sheep as a hobby—actually sheep that Jan had before she and Bill were married. They had worked up to a small flock of about 20. One night, Jan went to a meeting held by Joan Snyder of Hollow Road Dairy. Snyder is one of the first dairy sheep producers in the United States. The meeting was intended to recruit producers. Hollow Road was looking for people to supply sheep milk so they could expand their cheese and yogurt operation.

Jan and Bill were expecting son Tim at the time and were thinking about ways to earn extra money so that Jan would only need to work part-time. The land was there and more pasture was available. They decided to try it—they put in the equipment and expanded the flock to a hundred.

An old chicken barn served as the milking parlor. Sheep dairying was so unfamiliar that the dairy licensing was under "food," not dairy. Sheep dairy licensing has only been in effect since 1993.

In their expansion, they bought, as Bill says, "Everyone else's problems." Footrot plagued the flock, and they spent many hours in the barn with ewes standing in footbaths. In addition, the shift to a grain-enriched diet for better milk production caused a lot of health problems for the animals.

"We lost a lot of ewes," Bill says. "It was very discouraging. We thought we were killing them. I think it was a combination of diet and stress from the move to a bigger flock. Life became very hectic, especially after Tim arrived. We were trying to milk in time to get ready for work by nine." But the problems pushed the Butler's toward some decisions.

"We weren't doing this the right way," Bill says. "It was a management thing. We started thinking that we needed to get serious and do this the right way or not do it at all."

In September of 1991, the Butlers took a trip that pushed their thinking further. Jan's father had been raised in Taylor, Wisconsin, and still had a farm there. That's where Jan and Bill and baby Tim went for a vacation. Toward the end of the vacation, Bill was sitting on the porch one afternoon while Tim napped, and Jan went for a bike ride. "I was getting caught up in the romance of this area, the country life," Bill says. "I was thinking, Wouldn't it be nice to have this life, to live like this all the time?" Jan came back from her bike ride and was straddling her bike in front of the porch. "You know what I've been thinking?" Jan asked. "Yeah, I do." Bill answered.

Their daydream became closer to reality when they got back to Albany and talked about their vacation with Hollow Road Dairy's Joan Snyder. She told them to think seriously about it. She had been looking for ways to expand the dairy's size and market and was willing to consider a partnership with the Butlers, using the Hollow Road name. The next summer they made a return visit to look at real estate. In June of 1994, they moved to their new home, with the culled flock from their Albany operation.

The Butlers' 280-acre farm is surrounded by the rolling pasture and crop ground that the area is famous for. The new milking parlor was built inside a cattle building. One room holds the actual parlor, with a suspended platform built from metal tubing and a plastic, slatted floor. The sheep come in from the shed outside on a gentle ramp and move down the platform to stanchions that open consecutively. A ramp at the other end lets sheep outside after they are milked. A separate, brightly lit room next to the parlor holds the stainless steel processing equipment for their product, yogurt. There is a small bulk tank and a pasteurizer, where the yogurt culture is also added. Then there is a filling machine, which automatically measures out 5 ounces of milk into plastic containers and seals the lids. The cups are then put on trays and put into warming units—actually the same equipment used for raising dough. The finished yogurt is then stored in a walk-in refrigerator.

Bill and Jan have a rough division of labor. Bill does most of the milking and caring for the animals, a big jump for someone raised in New York City. "My family thinks this is pretty bizarre," he admits with a grin. Jan does the processing and manages the deliveries. Getting the equipment together and passing inspection took almost a year.

"The equipment was so small that we had a hard time piecing together everything we needed. But the companies were great," Bill says. "If they couldn't do it, they would send us to someone else. Working with the inspector in Madison was hard, but it was worthwhile. Every time he said no, he would explain that what he was doing was going to make it easier for us to have a clean, quality product."

The product is definitely yogurt, but it's different. It is smoother and richer tasting without the sour bite that store-bought plain yogurt has. It is to regular yogurt what Ben and Jerry's is to generic ice cream.

"Our product has no additives or stabilizers," Jan explains. "So extensive travel could hurt it. That has meant cultivating local markets in the health food sector."

Three health food stores in Madison are carrying the Hollow Road Farm yogurt now, as well as some stores in Milwaukee and Minneapolis.

**11.5.**  Jan and Bill Butler in their special milking parlor.

The farmers' market in Madison was a great success this summer, both from a sales standpoint and in letting people know who they are and promoting their product.

Getting people over some biases about their product has been an important part of the promotional activities. Jan and Bill smile at the recollection of their summer at the huge outdoor market in Madison. "You can read their lips," Jan says. "As they walk away, they're saying, 'Sheep?' Not everyone likes yogurt, but this isn't about the yogurt. It's about the sheep. It is so unfamiliar to them. I don't know what it is, whether they think it's dirty or what." On the other hand, the Madison farmers' market put them in touch with another group of people who understood their product completely.

"Madison is a very cosmopolitan city, with the university and all," Bill explains. "Many people would come back and say, 'I haven't tasted this since the last time I was in Greece.' Everyone who tries it says it tastes like what they have had in other countries, and they like it. They eat it there for breakfast. Two women came up who hadn't had it since they were little girls. They had been raised on an island in the Adriatic and their father had milked sheep and made sheep-milk yogurt. They loved it."

From their hectic beginnings, the Butlers are starting to plan for a future that they hope will eventually make the sheep dairy their only business, with no need for outside jobs. One part of this plan is starting to improve the production abilities of their flock. They would like to increase the 170 ewes they have now to 200, but with greater producing ability. Currently their "Heinz 57" mix of ewes is producing about 2 pounds of milk a day each, over the 4 or 5 months that they can milk. They have begun to add some Arcot Rideau stock, a breed developed in Canada using some of the heavy-producing East Freisian genetics. They have a ram and two ewes now and are hoping to add more through artificial insemination. Next year they will start to seriously cull, based on age and productivity. Right now, they say, any warm body will do.

"We've heard that some of the heavier-milking breeds can produce 4 to 5 pounds of milk a day over a 6-month period. From what we've seen of our Arcot Rideaus, we think that's probably right. But if we produce that much more, we're going to have to find new markets, too."

Jan thinks their next step in terms of product will be to expand the yogurt line. Right now because of the equipment, they only have maple and plain. Adding fruit and perhaps a low-fat product would give them a greater range of customers.

In addition to the genetics, nutrition will be playing an increasing role in their management plan. By crop sharing with a neighbor, they have access to all the hay they need without having to invest in hay equipment. The past 2 years they have used haylage, one silo and two 100-foot bags, in addition to square bales. They are beginning to purchase premixed concentrates through a local feed dealership.

"We are learning together how to do this," says Bill of his feed salesman. "He's never done this before either. We're experimenting, but what I've seen so far has been great—more milk over a longer period of time."

Likewise, veterinarians in the area have little experience with sheep, let alone dairy sheep. "We've had a great relationship with Cindy Wolf at the University of Minnesota veterinary school. She's brought a class out here and has also worked with us on a mastitis problem. She has done quite a bit of research with dairy sheep."

In cultivating their markets, the Butlers have realized that a big part of the barrier is getting a wider number of people comfortable with the idea of sheep dairy products.

"When we set up at the farmers' market we take along pictures of the farm and of the milking parlor," Jan says. "Some of the business, like the drive to Milwaukee, is not paying off financially yet. But we need to be there, to let people know we're here and that we have a good product."

Although they are not living in a yogurt-eating community, Bill and Jan have also gone to the farmers' market in Whitehall, 4 miles from their home.

"It's not a big market, but we've met people and have a chance to talk to the local people about what we're doing," Bill says.

"And," Jan adds, "we actually have cultivated some loyal customers here in Whitehall."

The Butlers are hopeful that tastes can change, even here in Holstein country.

## Niche Marketing in the Fiber Circuit

Randy Irwin of Freeland, Maryland, walks around with a piece of fleece in his jacket pocket at all times, ready to show weavers and spinners. "I say, 'What a beautiful piece you've made here! What kind of wool did you use?' And then I say, 'Here, touch this. Feel this. Wouldn't you like to make your spinning job easier?' I'll carry something that has some length, and maybe in a beautiful gray or silver or brownish color. That's marketing."

And there he is in an advertisement in the Maryland Sheep and Wool Festival Catalog, on page 31 of 120 pages, with a show picture of two of his prize-winning Border Leicester sheep. The world depicted in this catalog is a mysterious one to many experienced sheep producers, let alone to the beginner. The registered sheep here are Romney and Scottish Blackface, Lincoln and Cotswold, Karakus and Jacob sheep. There are listings for spinning equipment and spinning and weaving workshops. The Maryland show is now the largest wool crafts show in the country—1,000 sheep and 700 fleeces compete every year in various classes; 250 exhibitors set up booths. There is a wool auction, a shearing contest, and a sheep-to-shawl contest. And people come to buy high-quality animals and fleeces to go into their wool projects.

David Green, Carroll County Extension Service Director, started the show 23 years ago. The idea for it came when he was managing the Maryland wool pool and found more and more people looking for high-quality fleeces to spin. The fiber crafts wool market has grown, and the show is now a national fixture.

The prices that producers can receive for quality fleeces of the types that spinners and crafts people want are intriguing for producers. Clean, long, and resilient fleeces are desirable. Exotic fleeces and colored fleeces have appeal to select markets, as well. Prices can range from $2 to $10 a pound, so motivation is high.

"People do crazy things with fiber," Irwin says. "They might use dog hair and children's hair and birds' nests and everything else. I just came back from a fiber fair in North Carolina and there were llamas and alpacas and angora rabbits and emu birds and sheep. I've been at this for a long time. It may be a niche market but it's definitely expanding."

The Maryland Sheep and Wool show has a special appeal to Irwin because it is so close to his home. "I can come here and pay for my yearly feed bill in 3 days, selling registered sheep and the wool and the hides. I'd sell the teeth if I thought anybody wanted them."

Randy Irwin is not a spinner or a craftsperson. He sees himself as a producer of raw materials for those activities. He has a college degree in animal science and is interested in genetics and the challenge of making the animals and the land complement each other. His operation includes his flock of 30 purebred Border Leicester sheep, some angora goats, a llama, miniature donkeys, a Christmas tree farm, and, as he puts it, "A partridge in a pear tree." His day job, managing the sheep farm for Johns Hopkins University, which is in Baltimore, is a further reflection of his interests.

He also has more than a passing interest in the competition that the shows provide. "I love to compete," he says. "I can't play basketball and football anymore because I'd get killed. But I can go to a showring and carry through on things I thought about a year or two before, and I enjoy that. That's the therapy for me."

But he needs to make a profit on his business, and he does this by knowing his customers and cultivating the market that they provide.

"The spinners range from hobbyists to people in business. Some people just use spinning for therapy. They'll sit down and spin and talk and look at TV, and it's therapy for them. Then there are those who do a bit more and make clothing and garments for their families, and dress their grandkids in sweaters they spun and knit themselves. And there are those who open up a shop and sell yarn and raw fleece and hand-made garments and so forth. I see all of them, and I try to market to all of them."

Irwin looks for new markets too. "I go to craft shows before Christmas that have absolutely nothing to do with sheep at all. I walk around and see what materials people use in their products and see if I can talk them into trying some of my fleeces or hides in their things. People will send me examples of the products they make, a scarf or a duster. I had a lady contact me. She and her husband are in one of those musket loading rifle groups where they dress up in deer hides, and she had a sheep hide that she sat on in the booth where she was selling jewelry. She had so many people ask her about buying the hide that she came to me and got some to sell. I sold 30 hides to her. That's great for me because I don't have to sit at home and wait for someone to leaf through this pile of hides. That's what I'm trying to do. Some people like to have a little shop there on the farm and sell things, but I don't have the time. My marketing scheme is to do as much as I can without my being there."

While Irwin may be cultivating a market for Santa Claus doll beard material instead of for lamb chops, the basics of sheep production apply, with some special care for the wool. Genetics, nutrition, and management practices are still the essential ingredients.

The Border Leicester breed that Irwin has been raising has advantages

**11.6.**   Randy Irwin and a prize-winning Border Leicester yearling.

for him both for the wool and meat they produce, and for their adaptation to the hilly farm where he lives. The breed produces a long, coarse wool that is very attractive to spinners. Coarse, as Irwin explains, is not about roughness but about the length of the crimp, which in the Leicesters is more wavy than finely crimped, like the Rambouillet. A long, soft, but coarse-grade wool is what the spinners seem to want now, and he keeps an eye out for rams that have these kinds of fleeces to use in breeding. The best are soft to the touch and take dye very well. In addition, he finds the breed to be good forage users, which is essential. "I can't afford to feed grain in the summer," he says.

The Border Leicesters are also a dual-purpose breed. Although Irwin concentrates on wool production, he still has more ram lambs every year than are needed for breeding sales, and these lambs need to grow for the slaughter lamb market. Because of Irwin's time away from the farm, he likes the easy lambing of the breed. "I also just like the way they look," Irwin acknowledges. "They are clean faced, alert, and stylish looking."

Nutrition and breeding go hand in hand, Irwin testifies. "You can feed an animal well, but if he doesn't have the right genetics, he isn't going to perform for you. I think there is a difference in feeding for wool. The wool breeds don't grow the same as the meat breeds. A lot of the nutrition does go into wool production instead of meat. They don't grow quite as fast. If you don't provide the right nutrition for both carcass and fleece, then you're shorting yourself. I feed a 14% protein feed before and after lambing. I generally shear before lambing, so I am feeding not only the ewe but a set of lambs and getting the wool to grow as well. A lot of people overfeed and the sheep get too fat, but you have to be careful and remember how much you are asking of the animal."

Irwin sees that good proactive management is the best way to ensure a quality wool product from his sheep. "If you want a fleece that will bring $2.00 a pound instead of 43 cents, then it has to be clean." This means things like keeping pastures clean of burrs and thistles and other weeds. "But I don't keep the sheep indoors," Irwin says. "I have found that that is actually worse." Keeping the animals wormed so they don't get diarrhea, which stains the wool, and putting down straw in shady areas where the sheep lie in the summer is also important to prevent manure stains. Irwin shears before lambing because he has found that the lambs make a mess out of their mothers' fleeces. "She lies down and they jump all over her—on and off, on and off. You can't sell a fleece after the lambs have played on it." Since Irwin's operation is a purebred one, he uses a late January lambing time. "If I get a few lambs by mid January, all the better. For my show competition, that means a few more pounds and an extra inch on the fleece."

"When I got started here when I moved to Maryland, I was interested in the natural-colored sheep. I wanted to better the quality of my own animals and market the wool for a little better money than the commercial market. This area of the country is really about hobby farming. In the East you don't have a choice. The big farms are all split up and the hobbyists greatly influence this part of the industry. We can't do it with numbers. The idea is to keep your numbers down and the quality up, and the more diverse the breed and the more desirable and unusual the products you have to sell, the better the price."

And this gets back to marketing for that special audience that is interested in a high-quality product for a special purpose—niche marketing. Randy Irwin does it by competing, by going to sheep and fiber shows and crafts shows, with his award-winning animals and fleeces. He advertises in the special interest publications. And in his pocket is always his low-cost and ever-present marketing tool, a beautiful and intriguing sample of the product his animals create.

## Sheep as Teachers

Tom and Sandy Clayman of Hutchinson, Kansas, had a somewhat unusual goal for their sheep operation. "Our main purpose was to raise our kids," Sandy says, "and we've used the sheep to help us do that." She pauses for a minute. It is late afternoon on the Saturday before the end of the Kansas State Fair, and it has been a hectic week. The four Claymans, including Michelle, 14, and Michael, almost 17, were all going at full pace. "Well," she smiles, "I'm not sure if Tom would quite agree with that."

"I really like the sheep," Tom says, with mock defensiveness. Still, it is pretty clear that the sheep are at the center of the Clayman family life. As a symbol of both peace and responsibility, the Claymans might march some of their tall, white purebred Montadales through the halls of Congress as an answer to the decline in "family values."

This nuclear family history actually began with showing sheep. Sandy and Tom met at the Ohio State Fair. Tom's family raised Columbias and Sandy's father had Montadales. They got married and kept on raising sheep, and Tom worked in town—"To support my habits," he says. "The sheep industry has been part of our lives forever."

Continuing the breed roots, the Claymans now have 30 purebred Montadale ewes and 12 Columbias. For the Clayman children, the project is not just about buying 4-H Club lambs and coming home with blue ribbons. "They've learned the entire industry," Sandy says. "They were three and five when they had their own animals to take care of, but they were involved before that. Actually, Michael learned to walk at the Utah National Ram Sale!" From breeding, health care, marketing, shearing, and sheep industry business, the Clayman children have just about done it all.

Getting information and making decisions seem to be the two skills that the sheep have taught the best. As early as second grade, they would be making decisions about when to call the veterinarian. "It doesn't bother me to leave them for a show during lambing time," Tom says. "They know how to do it, and what they don't know, they know how to find out. Just get on the phone and get help."

Among the complexities that the Claymans deal with are the special issues of the purebred breeder. "Our focus has always been on a sheep that would work for the commercial producer and the purebred people, and sometimes that's a hard mix to keep. The commercial producer likes a thick make kind of sheep, an easy-to-keep kind of sheep. And the purebred industry likes 'em big and tall and long, with 30 miles of daylight underneath them. It's hard to keep a good, solid mix going. The sheep may look pretty, but is it productive? If you go around to purebred dealers and ask how many of their sales go to commercial breeders and how many to

**11.7.** The Clayman family portrait. *Left to right:* Sandy, Tom, Michael, Michelle.

the purebred industry, most definitely go to the purebred industry. That's kind of too bad. In regard to the wether and club lambs, we are working on carcasses that yield 58 or 59% instead of 40 to 50%. These are the genetics that the commercial person needs, but they don't know we have them."

Genetics and nutrition are the areas that the Claymans emphasize in their own operation and in their teaching to others in the industry.

"The biggest mistake people make in shows is not selecting the right genetics," Tom says. "You have to have a sheep that can walk and move freely, so they can be a foraging animal. And people normally don't take time to feed them correctly. Nutrition and genetics. You can trim and shape but you have to know genetics. My best advice to anyone getting into the sheep business is to get 10 half-sisters from the same ram, or better yet, get mother-daughter groups. If your genetics are alike at the beginning, then when you go to buy a ram, you can pick one to work with those genetics." The genetic traits that most affect productivity are feed conversion, lambing rates, and sexual maturity.

"With some of the emphasis on tall sheep," Tom says. "We are losing our feed conversion ability. When you put on so much structure, instead of being 4–5 pounds of feed for a pound of gain, you could be talking about 7 or 8. That's not good. We also want to push the sexual maturity a little. We look for rams that were born in January, so we know there is some propensity for the early lambs that we need in this business."

Another piece of advice from the Claymans has to do with feed. Beginners need to analyze potential grazing and match forages to the part of the country they are in.

"What works in Iowa or Nebraska is different than here. We're on the edge of being warm all winter, where 100 miles north it's different. Carrying feed buckets is very expensive. Here, if we run things right, we can have the sheep grazing except when there's snow on the ground, if we use bromes and fescues. In this part of the world, we can use wheat stubble, from November 1 to April 15. We supplement but we try to use a forage base all year long." They also recommend tapping into local feed companies. "In Oregon, people use cull onions. Turnips and peanut skins can be plentiful, too. Screenings from wheat or soybeans can be inexpensive in some areas. In South Dakota it might be sunflower meal. Corn screenings are sometimes the best part of the corn," Tom says. "If you can build your rations around something that costs 2 cents a pound, it helps a lot."

Wool has been a more significant part of the Claymans' operation than for many Midwest operations. For starters, Tom worked in the wool-buying business, so his interest in the quality and marketing of wool is high. "Wool is kind of the forgotten commodity. People need to understand that all fibers have their cycles. Right now the market is pretty good, but we're beginning to see some consumer resistance. I don't mean consumers at the stores, but our consumers at the mills."

Among the many aspects of the industry that the Clayman children have been involved in is wool and shearing. Michael has been shearing his own sheep now for 2 years, and he shears for neighbors as well, with Michelle catching sheep for him. "Michael was involved in a showmanship class a while ago," Tom says. "And the judge thought that the participants should know something about the industry. When the judge asked them 'What is the wool worth?' Michael had all kinds of follow-up questions: Is it coarse? A blackface? Fine wool? None of the other kids in the ring knew about that. I have had Michael call mills around the country to find out what they were paying for wool."

Knowing the product and knowing what you are going to do with it in advance is a critical part of beginning management that the Claymans see many people overlooking. "I could go to the sale barn on any day and find a hundred lambs that are there just because someone decided this was the day to get rid of them and take them off a feed bill. It's the difference between being a seller and a merchandiser. If he was a merchandiser, he would know he could feed that lamb for another 30 days and make money. Its the same with wool. The shearer comes and shears the wool. The person says, 'What's it worth? OK, I'll take it,' with no consideration of what the options might be. There isn't a major corporation in America that doesn't have someplace to go with their products. Even if you have just a few sheep, you need to have someplace to go with them."

As merchandisers, the Clayman children have developed their own market with a butcher shop in the area that sells high-quality meats. A small kill facility processes it, and the market pays them by live weight. They have arrangements never to go below 70 cents a pound, but never over 85, so there is some security at both ends.

"It's really been good for the kids to see the product in the meat counter case," Sandy says. "They can ask themselves: Are we doing this correctly? Is this the right kind of product?"

So now it is Sunday morning, the last day of the fair, and the Claymans, all four of them, are at the promotion booth for the Kansas Lamb Council in the Pride of Kansas pavilion, where everything Kansas is promoted, from soybeans to emus. The booth has literature on lamb, recipes, and other sheep products. Michelle is at a small electric grill, flipping lamburger patties for a tasting plate, and her father is standing by, talking to a visitor about the value of the product.

"As far as either Michelle or Michael being in the sheep business the way Sandy and I have been, at this time I don't really see it, and that doesn't bother me," Tom says. "It's been a great learning tool for them, teaching them responsibility—getting up, doing chores, taking care of things. Murphy's law is that when you don't check, something goes wrong, and they've both learned that. They are both better people for this. It's nice to be at the top, which they've been, but they've been at the bottom too, and that's a good lesson as well. You take your knocks."

Mothering ability may be one of the genetic traits of concern in the Claymans' purebred flock, but teaching ability is more to the point. It is clear that the Claymans' sheep have been successfully bred for this gene.

## The Bad Gene

There are lots of good genes at High Hill Farm. At the top of the namesake hill, a flock of purebred Suffolks grazing in tall golden grass shows off the genes, lifting their black heads in unison as a car drives by, prominent ears parallel to the ground. But there is a "bad" gene that owner Mary Ann Nipp of Arlington, Nebraska, says she has herself. None of her five grown children have it, she explains. "But," she says "I can spot it. When a kid comes with their family to pick out a club lamb and walks into the barn, I can spot right away the one who is going to take to it, the one who can't live right without being around livestock." That's the gene that Mary Ann Nipp has. In the peculiar and intense world of purebred sheep, Mary Ann's business and daily responsibilities are infused with a gentle passion for both the animals and the land that they live on.

Mary Ann began raising sheep in 1978. "I came from a cattle family in Idaho, but I needed a livestock project I could manage myself, so I tried sheep. I started with a small flock of 20 commercial ewes. Those dear old girls!" she says fondly. "They taught me a lot. They were old western ewes. You'd get in the jugs with them during lambing and they'd about jump out because they hadn't been around people before."

After a few years of learning, Mary Ann bought a small flock of Suffolks. "I thought, well, if I'm going to do this, it won't take a lot more time or money to have registered sheep. We have 300 ewes now, but the female base is pretty much from those original bloodlines."

While genetics are important to everyone who raises sheep, it takes on special significance for the purebred breeder. The purebred breeder has an eye for three main groups of customers. The first includes those who want to raise and/or show purebred sheep themselves. These lambs need to conform to the breed standards of phenotype, or appearance. Another audience is the club lamb market, the program for youth where

**11.8.** Mary Ann Nipp tracks the good genes on her Arlington, Nebraska, farm.

lambs are judged more on the basis of carcass. Most of the lambs shown as club lambs are purebred sheep, but there may be several breeds in a class, and it is the quality of the carcass that is judged. The club lamb shows do not require papers on the sheep. A third is the commercial customer who wants to use purebred genetics to improve the productivity of a flock.

One genetic concern in the purebred industry is what Mary Ann refers to as "tightly wound" or line breeding, which she has in her own long ewe families. "You have to be very careful with line breeding, and you have to know where your strengths and weaknesses are so you can capitalize on the strengths. There is something to the idea of hybrid vigor, even when it is just outside the line, not outside the breed. The lambs that are very tightly wound genetically are not as vigorous—they are slow when they're born. When you do an outcrossing of some kind, there's a sudden jump in vigor. I don't know why it works, but it does."

Mary Ann has some strong feelings about the purebred industry. "I don't feel any breed of livestock, purebred livestock, can be viable unless there is a strong commercial base. They have to produce a valuable product, there has to be a purpose. If we raise sheep that don't fit into that, I think we're kidding ourselves."

The challenge, in the showring and on the farm, can be the conflicting job of developing a phenotype of a sheep, a pretty sheep, that is of showring quality and also developing a sheep that contributes to commercial production.

"What the showring wants is a tall sheep, with a beautiful head, long bell ears, and real black points. Well," Mary Ann says with a smile, "you don't eat her head. The Suffolk is supposed to be a meat animal, but I have to consider the showring because that's where we sell a lot of our sheep." In fact, Mary Ann only sells 10–15% of her sheep to commercial operations. Everything else goes for club lambs or breeding stock.

Nipp has been on the board of the National Suffolk Association for 7 years. "At our last meeting, we started the process of redoing breed standards," she says. "There is a feeling that we are not getting enough of what we call depth of rib. We have been breeding longer sheep with no girth—that means the top of the back to the depth of the rib. We have leaned toward what I call a pistol-gutted sheep. The carrying capacity for the ewe, both in terms of the rumen (the ability to eat) and the rib cage (for ability to carry lambs), isn't good on those sheep. If you have a ewe with a deep rib cage, you know she can carry twins."

In some ways, Mary Ann, who also raises some cattle now, compares what is going on in the sheep industry with what went on with cattle earlier. "The industry can use big lambs. We want a bigger chop and loin, but we need sheep that are efficient too." But the sheep industry is different from the cattle industry in that there are many more 'hobby' producers in the sheep industry, and the economic factors aren't always as critical for them as they are for the commercial producer.

The words "do-ability" and "ease-of-keeping" are phrases that are used frequently in Mary Ann's conversation. They have to do with the ability of sheep to sustain themselves on forage and raise lambs without a lot of extra feeding.

Recordkeeping is the only way to control for the traits that affect productivity. Nipp has used Flock Expected Progeny Differences, a program developed by the National Sheep Improvement Program (see Chapter 4), which allows her to measure across her own flock. The traits that the program looks at are mothering ability, milking ability (which is measured by the weight of the lamb at weaning), and lamb weight gain. This year she started with the EPD program, Expected Progeny Differences, which will allow comparisons of performance across the breed, not just within her own flock.

"We are where the cattle industry was maybe 30 years ago in terms of developing objective measures of genetic performance," Nipp says. "I think it is going to take a long time before this is used very much. But I think people will eventually see that this is a tool for helping them make more money, make it more of a business. Right now in the cattle industry you can buy cattle confidently, just by the numbers."

But show business is a different business. You can have all the right genetic numbers, Nipp points out, and still not have a sheep that is aesthetically pleasing in the showring. Not pretty enough. "So, you walk the line as a breeder all the time," she says.

While Mary Ann takes a more scientific approach to her breeding program than many, she is still a salesperson and a counselor to the family that comes to her looking for lambs for their children. "People start calling in the spring and say that they want to start a small flock for their kids so they can show. And I have to ask them whether they want a breeding sheep or club lambs. I recommend the club lambs a lot. The reality is that it's a lot easier project for children. The fitting of a breed sheep, the preparation for the showring, is hard. You have to wash the sheep and card and do hand clipping. I do it, but I'm very impatient with it. With the club lambs you just have to slick shear them, just shear them as close as you can, and a child can do that. It might take them all afternoon, but they can do it—and they don't have to keep up with the paperwork. It can also be just a summer project. After the shows, the lambs can go to market."

Feeding and exercising show sheep is an important part of preparation. Sheep can lose condition very quickly, even just on a trip to a fair. "They lose their top, and they feel kind of mushy. The kid has to learn to feed the sheep the right way so they don't lose condition, but also so they don't get a big gut. Everyone in the showring thinks they have the right system for feeding," Mary Ann says. "You can give the kids some pointers, but everyone comes up with their own plan."

Maybe second to flock genetics, the topic in the industry is the ethics of showing sheep. Competition in the showring, for both club lambs and breeding sheep, can be strong. The National Suffolk Association puts on a large show every year to promote youth in agriculture. Eight hundred to 900 head of sheep come with children and their families from all over the country.

"We have strong rules and guidelines in the National Suffolk Association," Mary Ann says. "And for every one of those people who gets a little too competitive, there are all these families having a great time. If you come to our National Junior Show, you'll see 500 kids—with 500 families and grandmas and grandpas and a van full of brothers and sisters and all their livestock. These kids have a good time. And if it's done right, they can learn so much. Whether the kids have my 'bad gene' or whether they are there because the family wants to teach them responsibility, it's fun either way. What I have seen is that people come in with high aspirations, and they think if they buy good stock they can go out and win. And it doesn't work that way. There is so much to learn about feeding a lamb and getting it ready for the showring. They have to be patient and learn and not get discouraged if they don't win the first year out."

Sheep are Mary Ann Nipp's main business, so she is clearly concerned about marketing her product. But the purebred industry is one where reputation may be as important as advertising. "In the purebred business, marketing can be a slow process," Mary Ann says. "A reputation for honesty builds slowly and people need to know that you'll stand behind your breeding program. You could probably do it faster, but not if you're in it for the long haul. Of course," she adds with a grin, "I have nightmares all winter with these lambs on the ground, nightmares that no one will show up to buy them in the spring!"

Her 16 years in the purebred business has brought a healthy share of awards in the showring, but it is not her first thought when she thinks of her success. "The most rewarding thing for me is to feel I've earned the respect of my peers. That has been very rewarding. I've always had a love of livestock, and being able to carry this dream out has been great."

The "bad gene," it turns out, has a lot of benefits. "I think people in the livestock industry are wonderful. You can go anyplace, and you know people and you're welcome and you believe in the same things. I think it's a great privilege to have livestock and land," she says. "It's a blessing, and I'm grateful for it everyday."

## The Most Important Equipment

It shouldn't be surprising that Stan and Jean Potratz of Washington, Iowa, have some answers to the question, "What advice do you have for beginners?" They have their own years of experience with commercial sheep production—15 years together and more for Stan, both in this country and in England. On top of that, they have the years of experience with their sheep supply business, Premier, where part of the job is answering questions and providing counseling to shepherds with problems. In addition to being shepherds, they are scientists and have experimented with many approaches to management as well as with the equipment they sell. Everyday, they are both teachers and students in the shepherding life.

The most important part of their answer is this: Buy disease-free animals of a breed, or breed mix, adapted to the climate. Plan management—feeding programs, lambing schedules, and health maintenance—around the particular environment you are in. The answer is that there is no one answer. What works in Colorado does not work in Iowa, does not work in South Carolina.

"We currently have about 400 head of ewes," Stan says. "The genetics are primarily Border Leicester. We will be moving to Border Leicester/Clun Forest or Border Leicester/Dorset in the future. We will be using a terminal sire of the blackface variety. This is a commercial flock. We don't sell breeding stock per se, although people could buy breeding stock from us if they wanted. We probably have one of the larger Border Leicester flocks this side of Oregon.

"We use this breed because we have learned several things over the last 15 years," Stan continues. "One is that most of the years here, the grass tends to be wet. The ordinary ewes that come out of the West—the Rambouillet-type whiteface ewes—have smaller stomachs adapted for arid conditions and dryer grasses. Here in Iowa, those ewes tend to fill up on water. They can't consume enough of the wet grasses to nourish themselves."

Part of Stan's background includes the more than 10 years he spent in England, both as a college student and then as manager of the college flock. "I noticed in the high wet hills of Scotland they have a coarse-wooled sheep called the Scottish Blackface. Same purpose. It has a very large belly relative to its frame, allowing it to consume a lot of wet grass. These Border Leicesters are like that. They have a bigger rumen than the western sheep and thus have more room in the stomach to feed themselves and the lambs."

Jean and Stan have always concentrated on making use of the ruminant abilities of the flock on the rolling hills of south-central Iowa. "We have a farm that isn't suitable for raising grain," Jean says. "We need a ruminant animal to take care of all the pasture."

"My opinion," says Stan, who has many strong ones, "is that the whiteface range sheep breeds, or even the Suffolk, which is bred to be trim for show purposes, aren't well adapted to conditions here. You can make it work if you are feeding a lot of grain. But if you use a lot of legumes, as we do, clovers in particular, which have an even higher moisture content than grasses, the sheep tend to practically starve to death on our pasture."

Another aspect of buying adapted animals has to do with disease control. "The animals we have now are all ours," Stan says. "We have only

**11.9.** Stan and Jean Potratz with some of their well-bred ewes.

brought in 8 rams in the last 12 years. The rest of them are ours. All 400 ewes were born here. We have dealt with all kinds of health problems in the flock, but now they are pretty much disease free—no footrot, no exotic diseases. The beginner doesn't have this luxury, but every time you buy sheep, you buy a certain set of problems that you have to deal with."

Jean confirms this. "Often the beginner will put together a flock by buying a few sheep here, a few from this farm, and a few from the sale barn. With every different set, you are buying a slightly different set of problems and resistances. At least if you buy from one farm, you are buying only one set of problems to deal with. You are much better off buying from a farmer too, rather than a sale barn. At least you can go see if they're walking around on their knees with footrot."

The Potratzes warn that a premium must be paid for the sheep they are recommending, maybe 40% more than for an ordinary sheep in a sale barn. They say to figure paying one and a quarter times what you would pay for a market ewe lamb. You may not understand why the premium is important until you've got sick animals.

"On the other hand," Jean says with a smile, "if you're going to make mistakes, inexpensive sheep are better as guinea pigs."

The management routines of their flock are also created for the particular climate and resources that they have. They used to lamb entirely outside with almost no buildings. Now they have reversed that situation. They have a large open indoor building, and the ewes are actually inside the building for almost half the year.

"This is a response to our climate here," Stan says. "The productivity of our pastures is central to us. We run 400 ewes on 120 acres, so that's pretty intensive." What Potratz found was that, because of the climate in this part of Iowa, there would not be one freeze that would last until spring, as there might be in Minnesota. Instead there were periods of thawing that would allow the sheep to wreak havoc on the pastures in the winter. Mud is common and the sheep suffer, and the grasses are cut up and ruined. "It doesn't apply to every midwestern farm, but it happens in this part of Iowa," Jean says.

The sheep graze on carefully managed pastures from April until the growing stops toward the end of October. Then the lambs are sold and the ewes come inside and don't leave again until the end of April with their new lambs. Each ewe has about 15 to 20 square feet of space inside, and they are kept in groups of 20. In addition to their ruminant capacities, the Border Leicesters are also a calm sheep, according to Stan, and can handle the constant human presence of an indoor facility.

The damp conditions of the Midwest and the heavily used pasture system that the Potratzes use make parasites a major problem and a focus of the shepherding calendar. "We worm the sheep every 3 weeks when they are on pasture in May and June," Stan says. "We do this religiously. We do it even if we have to stop making hay to do it. We have found that 1 day over and they start reinfecting the pastures."

Stan and Jean have done a great deal of thinking about the profitability of their flock, and the special problems in trying to make a living on sheep in the Midwest.

"There are 6,500 flocks in Iowa," Stan says. "But I estimate less than 10 of them produce over half the owner's gross income from the sale of fat lambs. This is a reduction from 12,000 flocks 15 years ago. Looking after sheep in small numbers is a lot of work. The primary people who profit are purebred breeders and ones who sell to spinners or 4-H lamb people.

These are people who have a specialized market, and they supply what that special market wants. But there are not many commercial producers."

Despite the emphasis on pasture in their program and their appreciation of the particular value of the ruminant, those particular strengths can't always be used in the Iowa economic situation. "Hay is relatively expensive here," Stan says, "And grain is relatively cheap. At least right now. So when the ewes are inside, we feed them the minimal amount of hay to keep the rumen going and the rest of their diet is grain. Iowans would be foolish not to take advantage of cheap grain while they have it," Stan says. "It's the one advantage we have in the market. The Midwest is just a more expensive place to raise sheep. Land prices as well as hay prices are high. We have no public land, so we are competing with producers in the West who have an advantage. You rarely see a commercial flock of any size in Iowa."

Why aren't there more lamb feedlots in Iowa, with all the inexpensive grain? One problem, according to the Potratzes, is access to feeder lambs. At least 1,000 lambs at a time are needed, a consistent group, and they are hard to find. The other thing is the weather. The rain and snow and mud make a mess of feedlots in the Midwest.

Making a midwestern commercial flock economically viable is a subject of much concern to the Potratzes, but as yet they have no simple solutions. They have, for one thing, a more concrete understanding of the labor costs than most midwestern farmers. This is because, with the demands of their equipment business, they have a full-time shepherd, and they know how much time he spends on their flock.

"One year he was gone and we had to do all the flock work ourselves, so we kept track," Stan says. "We were appalled. We figured we were putting an $8 bill for labor on each market lamb, just during the month of lambing." At levels of 50–400 ewes, the labor is manageable. But 1000 ewes might be what is needed to turn a profit, and they haven't figured out how to profitably manage that many sheep yet.

The Potratzes also think the midwestern producers may need to develop some of their own products for a regional appeal, and not compete on the national market with lambs that western producers can grow so much cheaper. The pork producers in Iowa have developed an Iowa Chop that has been successful. Something with similar regional appeal is needed for lamb as well.

Despite their Premier catalog business, the Potratzes' recommendations for equipment for the beginner are pretty minimal. "For a one-ram flock of maybe 40, which is a good place for the beginner to start," Jean says, "you don't need much. Feeders are important, but they don't have to be fancy."

"The feeders are to keep the shepherd away from the sheep," Stan says. "Walking through a flock with buckets can be dangerous. Kids will be more inclined to help with an operation if they don't get knocked around. That means fenceline feeders. They also must be feeders that keep the feces and the feed separate to help with parasite control."

Gates and panels are also important equipment, lots of them, especially for lambing pens and to crowd sheep into small areas for treatment. Equipment for drenching and injections is important as well.

"And, a Pritchard teat," Jean adds. "I wouldn't go through lambing without one of those. Also, stomach 'tubers' for feeding weak lambs, and a lambing bucket if you are going to have more than two orphan lambs."

Another important piece of equipment can be a guard dog. The Potratzes raise guard dogs for sale. Their dogs have done a good job of guarding their property so that the coyotes have basically gone elsewhere to feed. "I don't really want the neighborhood coyotes killed," Stan says with a smile. "Then we'd get a whole new pack of animals in that haven't been trained by our dogs." They haven't had a coyote kill in a decade, even though they can hear them singing in the hills.

"One problem with the guard dogs is feeding them," Jean says. "People don't realize what a hassle it is to feed the dogs with the sheep. The sheep eat the dog food." The Potratzes' dogs are actually not in with the sheep but in a pasture adjacent to them. This seems to be enough of a threat to deter the coyotes.

A final word of advice for the beginner comes from Stan, a remnant phrase from his years in England. "Don't be too clever by half," he says. "That means keep it simple. Start with the simple way of doing things. Lamb once a year at one time—don't try for accelerated lambing. Don't get exotic breeds that no one around you has ever raised. These things are for the very experienced shepherd."

And back to the equipment.

"The most important piece of equipment is the first sheep you buy," Stan says. "Buying the wrong kind of sheep, or buying diseased animals, can make your life miserable."

## REFERENCES

1. Railton, Arthur. "The Indians and the English on Martha's Vineyard," *The Dukes County Intelligencer*, February 1993, Vol 34, No. 3.

2. Freeman, James. "Dukes County—1807." *The Dukes County Intelligencer,* The Dukes County Historical Society, Vol. 12, No. 4, May 1971.

3. Klah, Hasteen. Navajo Creation Myths: The Story of the Emergence, recorded by Mary Wheelwright. Santa Fe, NM: Museum of Navajo Ceremonial Art, 1942.

4. Bailey, Lynn. *If You Take My Sheep, I Will Die.* Pasadena: Westernlore, 1980, p. 75.

5. Witherspoon, Gary. "Sheep in Navajo Culture and Social Organization," *American Anthropologist,* Vol. 75, No. 5, 1973.

6. US Department of Agriculture Economic Research Service, 1995.

# APPENDIX A:
## State Extension Service Contacts

**ALABAMA**
Extension Sheep Specialist
Auburn University
Animal/Dairy Sciences Department
Auburn, AL 36849
205-844-1520

**ALASKA**
Extension Livestock Specialist
University of Alaska
Fairbanks, AK 99775-5200
907-474-6357

**ARIZONA**
University of Arizona
Animal Science Department
Tucson, AZ 85721
602-621-9757

**ARKANSAS**
Extension Livestock Specialist
University of Arkansas
PO Box 39
Little Rock, AR 72203
501-671-2180

**CALIFORNIA**
Extension Animal Nutritionist
University of California
Davis, CA 95616
916-752-0525

**COLORADO**
Extension Sheep/Wool Specialist
Colorado State University
Animal Science Department
Ft. Collins, CO 80523
303-491-1321

**CONNECTICUT**
University of Connecticut
Animal Science Department
U-40 3636 Horsebarn Road
Storrs, CT 06269-4040
203-486-2636

**DELAWARE**
Extension Livestock Specialist
University of Delaware
Kent City Extension
Dover, DE 19901
302-697-4000

**FLORIDA**
Extension Livestock Specialist
University of Florida
Route 3, Box 4370
Quincy, FL 32351
904-627-9236

**GEORGIA**
Extension Animal Science Department
University of Georgia
Department of Animal Science
Athens, GA 30602
706-542-2875

**HAWAII**
Extension Specialist
University of Hawaii
Animal Science Department
1800 East/West Road
Honolulu, HI 96822
808-956-7090

**IDAHO**
Extension Animal Scientist
University of Idaho
Extension Livestock Programs
PO Box 29
Soda Springs, ID 83276
208-547-4354

**ILLINOIS**
Extension Sheep Specialist
University of Illinois
321 Mumford Hall
1301 W. Gregory Drive
Urbana, IL 61801
217-333-7351

**IOWA**
Extension Livestock Specialist
Iowa State University
109 Kildee Hall
Ames, IA 50011
515-294-5247

**KANSAS**
Extension Sheep Specialist
Kansas State University
Animal Science Department
Manhattan, KA 66506-0201
913-532-5790

**KENTUCKY**
Extension Sheep Specialist
University of Kentucky
Lexington, KY 40546-0215
606-257-2716

**LOUISIANA**
Louisiana State University
Animal Science Department
Room 204 Knapp Hall
Baton Rouge, LA 70803
504-388-6702

**MAINE**
Livestock Specialist
University of Maine
338 Hitchner Hall
Orono, ME 04469-0163
207-581-2789

**MARYLAND**
Extension Livestock Specialist
University of Maryland
Animal Science Department
College Park, MD 20742
301-405-1394

**MASSACHUSETTS**
Extension Livestock Specialist
University of Massachusetts
2 Hatch Lab
Amherst, MA 01003
413-545-2573

**MICHIGAN**
Extension Sheep Specialist
Michigan State University

Animal Science Department
113 Anthony Hall
East Lansing, MI 48824
517-336-1388

**MINNESOTA**
Extension Animal Scientist
University of Minnesota
1404 Gortner Avenue
St. Paul, MN 55108
612-624-0766

**MISSISSIPPI**
Mississippi State University
Box 5446
Mississippi State, MS 39762
601-325-3515

**MISSOURI**
Sheep Specialist
Sheep/Goat/Small Livestock Specialist
Lincoln University
900 Moreau Drive
Jefferson City, MO 65101
314-681-5533

**MONTANA**
Extension Sheep Specialist
Montana State University
221 Linfield Hall
Bozeman, MT 58717
406-994-3414

**NEBRASKA**
Extension Sheep Specialist
University of Nebraska
204C Animal Science
Lincoln, NE 68583-0908
402-472-6433

**NEW HAMPSHIRE**
Area Extension Specialist
University of New Hampshire
Route 1, Box 99
Westmoreland, NH 03467
603-352-4550

**NEW JERSEY**
Sheep Specialist
Rutgers University/Cook College
PO Box 231

New Brunswick, NJ 08903
908-932-9514

**NEW MEXICO**
Extension Sheep Specialist
New Mexico State University
Animal Science Department
Box 3AE
Las Cruces, NM 88003
505-646-1318

**NEW YORK**
Sheep Specialist
Cornell University
Animal Science Department
255 Morrison Hall
Ithaca, NY 14853
607-255-2869

**NORTH CAROLINA**
Extension Animal Specialist
North Carolina State University
Box 7621
Raleigh, NC 27695-7621
919-515-2761

**NORTH DAKOTA**
Extension Sheep Specialist
North Dakota University
Animal Science Department
Hultz Hall
Fargo, ND 58105
701-237-7645

**OHIO**
Sheep Extension Specialist
Ohio State University
2029 Fyffe Road
Columbus, OH 43210
614-292-6791

**OKLAHOMA**
Extension Sheep Specialist
Oklahoma State University
Animal Science Department
109 Animal Science Building
Stillwater, OK 74078
405-744-6065

**OREGON**
Extension Sheep Specialist
Oregon State University

Animal Science Department
214 Withycomb Hall
Corvallis, OR 97331-6702
503-737-4926

**PENNSYLVANIA**
Extension Sheep Specialist
Pennsylvania State University
316 Henning Building
University Park, PA 16802
814-863-3669

**RHODE ISLAND**
Sheep Specialist
University of Rhode Island
Fisheries/Animal/Vet Science Department
Kingston, RI 02881-0804
401-792-4183

**SOUTH CAROLINA**
Clemson University
Animal Science Department
Clemson, SC 9634-0361
803-656-5156

**SOUTH DAKOTA**
Sheep Specialist
South Dakota State University
Animal/Range Science Department
Box 2170
Brookings, SD 57007-0392
605-688-5433

**TENNESSEE**
Extension Livestock Specialist
University of Tennessee
PO Box 110019
Nashville, TN 37222-0019
615-832-8341

**TEXAS**
Extension Sheep Specialist
Texas A&M University
Research and Extension Center
7887 North Highway 87
San Angelo, TX 76901
915-653-4576

**UTAH**
Area Livestock Specialist
Sevier County Extension Service

Utah State University
250 North Main
Richfield, UT 84701
801-896-4609

**VERMONT**
Sheep Specialist
University of Vermont
Cooperative Extension Service
W. Main Street, Box 624
Newport, VT 05855
802-656-2992

**VIRGINIA**
Extension Sheep Specialist
Virginia Polytechnic Institute
Animal Science Department
Blacksburg, VA 24061-0306
703-231-9159

**WASHINGTON**
Extension Veterinarian
Washington State University
126 Clark Hall
Pullman, WA 99164-6310
509-335-2881

**WEST VIRGINIA**
Extension Sheep Specialist
West Virginia University

115 West Court
Kingwood, WV 26537
304-329-1391

**WISCONSIN**
Extension Livestock Specialist
University of Wisconsin
Meat/Animal Science Department
#438 Animal Science Building
1675 Observatory Drive
Madison, WI 53706-1284
608-263-4300

**WYOMING**
Sheep/Wool Extension Specialist
University of Wyoming
College of Agriculture
Animal Science Department
Box 3354 University Station
Laramie, WY 82071
307-766-2364

**USDA**
National Program Leader
Extension Services USDA
Room 3334, South Building
Washington, DC 20250
202-720-2677

# APPENDIX B:
## Sheep Supply Catalog Sources

**MID-STATES LIVESTOCK SUPPLIES**
125 East 10th Avenue
Hutchinson, KS 67505
316-663-5147

**PIPESTONE VETERINARY SUPPLY**
1300 South Highway 75
PO Box 188
Pipestone, MN 56164
1-800-658-2523

**PREMIER**
Box 895
Washington, IA 52353
1-800-282-6631

**SHEEPMAN SUPPLY**
14502 Old Gardonsrule Rd.
Orange, VA 22960
1-800-336-3005

**SYDELL, INC.**
Rt. 1, Box 85
Burbank, SD 57010
1-800-842-1369

**WOOLTIQUE INCORPORATED**
P.O. Box 537
1111 Elm Grove St.
Elm Grove, WI 53122-0537
414-784-3980

# INDEX